Pierre Joseph van Beneden

Animal parasites and messmates

Pierre Joseph van Beneden

Animal parasites and messmates

ISBN/EAN: 9783337228750

Printed in Europe, USA, Canada, Australia, Japan

Cover: Foto ©berggeist007 / pixelio.de

More available books at **www.hansebooks.com**

THE INTERNATIONAL SCIENTIFIC SERIES.

ANIMAL PARASITES AND MESSMATES.

BY

P. J. VAN BENEDEN,

PROFESSOR AT THE UNIVERSITY OF LOUVAIN, CORRESPONDENT OF THE INSTITUTE OF
FRANCE.

WITH EIGHTY-THREE ILLUSTRATIONS.

NEW YORK:

D. APPLETON AND COMPANY,

1, 3, AND 5 BOND STREET.

1887.

CONTENTS.

INTRODUCTION.

CHAPTER I.
ANIMAL MESSMATES.

CHAPTER II.
FREE MESSMATES.

CHAPTER III.
IXED MESSMATES.

CHAPTER IX.

PARASITES THAT MIGRATE AND UNDERGO METAMORPHOSES.

CHAPTER X.

PARASITES DURING THEIR WHOLE LIFE.

LIST OF ILLUSTRATIONS.

INTRODUCTION.

" The edifice of the world is only sustained by the impulses of hunger and love."—SCHILLER.

In that great drama which we call Nature, each animal plays its especial part, and He who has adjusted and regulated everything in its due order and proportion, watches with as much care over the preservation of the most repulsive insect, as over the young brood of the most brilliant bird. Each, as it comes into the world, thoroughly knows its part, and plays it the better because it is more free to obey the dictates of its instinct. There presides over this great drama of life a law as harmonious as that which regulates the movements of the heavenly bodies; and if death carries off from the scene every hour myriads of living creatures, each hour life causes new legions to rise up in order to replace them. It is a whirlwind of being, a chain without end.

This is now more fully known; whatever the animal may be, whether that which occupies the highest or the lowest place in the scale of creation, it consumes water and carbon, and albumen sustains its vital force.

Therefore, the Hand which has brought the world out of chaos, has varied the nature of this food; it has proportioned this universal nourishment to the necessities and the peculiar organization of the various species which have to derive from it the power of motion and the continuance of their lives.

The study whose aim is to make us acquainted with the kind of food adapted to each animal constitutes an interesting branch of Natural History. The bill of fare of every animal is written beforehand in indelible characters on each specific type; and these characters are less difficult for the naturalist to decipher than are palimpsests for the archæologist.

Under the form of bones or scales, of feathers or shells, they show themselves in the digestive organs. It is by paying, not domiciliary, but stomachic visits, that we must be initiated into the details of this domestic economy. The bill of fare of fossil animals, though written in characters less distinct and complete, can still be very frequently read in the substance of their coprolites. We do not despair even to find some day the fishes and the crustaceans which were chased by the plesiosaurs and the ichthyosaurs, and to discover some parasitic worms which had entered with them into the convolutions of the intestines of the saurians.

Naturalists have not always studied with sufficient care the correspondence which exists between the animal and its food, although it supplies the student with information of a very valuable kind. In fact, every organized body, whether conferva or moss, insect or mammal, becomes the prey of some animal; every organic substance, sap or blood, horn or feather, flesh or bone, disappears under the teeth of some one or other of these; and to each kind of *débris* correspond the instruments suitable for its assimilation. These primary relations between living beings and their alimentary regimen call forth the activity of every species.

We find, on closer examination, more than one analogy between the animal world and human society; and without much careful scrutiny, we may say that there is no social position which has not (if I may dare to use the expression) its counterpart among the lower animals.

The greater part of these live peaceably on the fruit of their labour, and carry on a trade by which they gain their livelihood; but by the side of these honest workers we find also some miserable wretches who cannot do without the assistance of their neighbours, and who establish themselves, some as *parasites* in their organs, others as *uninvited guests*, by the side of the booty which they have gained.

Some years ago, one of our learned and ingenious colleagues at the University of Utrecht, Professor Harting, wrote a charming book on the industry of animals, and demonstrated that almost every trade is known in the animal kingdom. We find among them miners, masons, carpenters, paper manufacturers, weavers, and we may even say lace-makers, all of whom work first for themselves, and afterwards for their progeny. Some dig the earth, construct and support vaults, clear away useless earth, and consolidate their works, like miners; others build huts or palaces according to all the rules of architecture; others know intuitively all the secrets of the manufacturers of paper, cardboard, woollen stuffs or

xvi   INTRODUCTION.

lace ; and their productions need not fear comparison
with the point-lace of Mechlin or of Brussels. Who has
not admired the ingenious construction of the beehive
or of the ant-hill, or the delicate and marvellous structure of the spider's web? The perfection of some of
these works is so great and so generally appreciated,
that when the astronomer requires for his telescope a
slender and delicate thread, he applies to a living shop,
to a simple spider. When the naturalist wishes to test
the comparative excellence of his microscope, or requires
a micrometer for infinitely little objects, he consults, not
a millimetre, divided and subdivided into a hundred or
a thousand parts, but the simple carapace of a diatom,
so small and indistinct that it is necessary to place a
hundred of them side by side to render them visible to
the naked eye : and still more, the best microscopes do
not always reveal all the delicacy of the designs which
decorate these Lilliputian frustules. Mons. H. Ph. Adan
has lately shown, with an artist's talent, the infinite
beauties which the microscope reveals in this invisible
world.

To whom do the manufacturers of Verviers or of
Lyons, of Ghent or of Manchester, apply for their raw
materials? Either to an animal or a plant; and even
up to the present time we have had sufficient modesty
not to have sought to imitate either wool or cotton. Yet
these animal manufacturers carry on their operations
every day under our eyes, the doors wide open to everybody, and none of them is as yet marked with the trite
expression, " No admittance."

" The beau-ideal which we place before us in the
arts of spinning and weaving," said an inhabitant of the

South to Michelet, "is the beautiful hair of a woman: the softest wool, the finest cotton, is very far from realizing it." The Southerner seemed to forget that this soft wool, as well as this fine cotton, was not the product of our manufacturers any more than the woman's hair.

Were these animal machines to sustain injury, or even to be idle for a certain time, we should be reduced to have nothing wherewith to cover our shoulders : the fine lady would have neither Cashmere shawl, silk, nor velvet in her wardrobe ; we should have neither flannel nor cloth to make our clothes ; the herdsman even would not have his goat's skin to protect him from the inclemency of the season. Thanks to the animal which gives us his flesh and his fleece, we are able to leave the southern regions, to brave the rigour of other climes, and establish ourselves side by side with the reindeer and the narwhal, in the midst of eternal snow.

We have our science and our steam-engines, of which we are justly proud ; the animals have only their simple instinct to enable them to fabricate their marvellous tissues, and yet they succeed better than ourselves. The so-called blind forces of nature produce thread, the use of which the genius of man seeks in vain to supersede ; and we do not even dream of entering into competition with these living machines which we daily crush under our feet.

All these occupations are openly carried on; and if there are some which are honest, it may be said that there are others which deserve another character. In the ancient as well as the new world, more than one animal resembles somewhat the sharper leading the life of a great nobleman; and it is not rare to find, by the side of the humble pickpocket, the audacious brigand of the high road, who lives solely on blood and carnage. A great proportion of these creatures always escape, either by cunning, by audacity, or by superior villany, from social retribution.

But side by side with these independent existences, there are a certain number which, without being parasites, cannot live without assistance, and which demand from their neighbours, sometimes only a resting-place in order to fish by their side, sometimes a place at their table, that they may partake with them of their daily food; we find some every day which used to be considered parasites, yet which by no means live at the expense of their hosts.

When a copepode crustacean instals himself in the pantry of an ascidian, and filches from him some dainty morsel, as it passes by; when a benevolent animal renders some service to his neighbour, either by keeping his rack clean, or removing detritus which clogs certain organs, this crustacean or this animal is no more a parasite than is he who cowers by the side of a vigilant and skilful neighbour, quietly takes his siesta, and is contented with the fragments which fall from the jaws of his companion. We may say the same thing of the fish which, through idleness, attaches itself, like the remora, to a neighbour who swims well, and fishes by his side without fatiguing his own fins.

The services of many of these are rewarded either in protection or in kind, and *mutuality* can well be exercised at the same time as *hospitality*.

Those creatures which merit the name of parasites

feed at the expense of a neighbour, either establishing themselves voluntarily in his organs, or quitting him after each meal, like the leech or the flea.

But when the larva of an ichneumon devours, organ after organ, the caterpillar which serves him as a nurse, and at last eats her entirely, can we call him a parasite? According to Lepelletier de Saint-Fargeau, who has so successfully treated these questions, the parasite is he who lives at the expense of another, eating that which belongs to him, but not devouring his nurse herself. Nor is the ichneumon a carnivorous animal, for the true beast of prey cares nothing at any period of his existence for the life of his victim.

True parasites are very commonly found in nature, and we should be wrong were we to consider that they all live a sad and monotonous life. Some among them are so active and vigilant that they sustain themselves during the greater part of their life, and only seek for assistance at certain determinate periods. They are not, as has been supposed, exceptional and strange beings, without any other organs than those of self-preservation. There is not, as was formerly supposed, a *class* of parasites, but all the classes of the animal kingdom include some among their inferior ranks.

We may divide them into different categories.

In the first of these we will place together all those which are free at the commencement of their life, which swim and take their sport without seeking assistance from others, until the infirmities of age compel them to retire into a place of refuge. They live at first like true Bohemians, and are certain of getting invalided at last in some well-arranged asylum. Sometimes both the male and female require this assistance at a certain age; with others it is the female only, as the male continues his wandering life. In some cases, the female carries her partner with her, and supports him entirely during his captivity; her host nourishes her, and she in her turn feeds her husband. We find few female gill-suckers which have not with them their Lilliputian males, which, like a shadow, never quit them. But we also find males, living as parasites of their females, among those curious crustaceans known by the name of cirrhipeds. All the parasitical crustaceans are placed in this first category.

We find others, the ichneumons for example, which are perfectly at liberty in their old age, but require protection while young. There are many of these, which as soon as they escape from the egg, are literally put out to nurse; but from the day when they cast off their larval robe, they are no longer under restraint, but, armed cap-à-pie, they rush eagerly in quest of adventure, and die like others on the high road. In this category are generally found parasitical hymenopterous and dipterous insects.

Other kinds are lodgers all their lives, though they change their hosts, not to say their establishment, accordingly to their age and constitution. As soon as they quit the egg, they seek for the favours of others, and all their itinerary is rigorously traced out for them beforehand. Fortunately we are at present acquainted with the halting-places and magazines of a great number of those which belong to the order of cestode and trematode worms. These flat and soft worms begin life usually as vagabonds, aided by a ciliary robe which serves as an apparatus for locomotion; but scarcely have they tried to use their delicate oars, before they demand assistance, lodge themselves in the body of the first host that they meet, whom they abandon for another living lair, and then condemn themselves to perpetual seclusion.

That which adds to the interest inspired by these feeble and timid beings is, that at each change of abode, they change also their costume; and that when they have reached the limit of their peregrinations, they assume the virile toga—we had almost said, the wedding robe. The sexes appear only under this later envelope; up to this period they have had no thoughts of the cares of a family. It has always been somewhat difficult to establish the identity of those persons who frequent the public saloons one day, and are found on the next in the most obscure haunts, dressed as mendicants. Most of the worms which have the form of a leaf or a tape give themselves up to these peregrinations, and those which do not arrive at their last stage, die usually without posterity.

It is interesting to remark that these parasitical worms do not inhabit the various organs of their neighbours indiscriminately, but all begin their life modestly in an almost inaccessible attic, and end it in large and spacious apartments. At their first appearance they think only of themselves, and are contented to lodge, as *scolices* or vesicular worms, in the connective tissue of the muscles, of the heart, of the lobes of the brain, or even in the ball of the eye; at a later stage, they think of the cares of a family, and occupy large vessels like the digestive or respiratory passages, always in free communication with the exterior; they have a horror of being enclosed, and the propagation of their species requires access to the outer air.

In the last category are found those which need assistance all their lives; as soon as they have penetrated into the body of their host, they never remove again, and the lodging which they have chosen serves them both as a cradle and a tomb.

Some years since, no one suspected that a parasite could live in any other animal than that in which it was discovered. All helminthologists, with few exceptions, looked upon worms in the interior of the body as formed without parents in the same organs which they occupy. Worms which are parasites of fish, had been seen a long time before this in the intestines of various birds: experiments had even been made to satisfy observers of the possibility of these creatures passing from one body to another; but all these experiments had only given a negative result, and the idea of inevitable transmigration was so completely unknown that Bremser, the first helminthologist of his age, raised the cry of heresy, when Rudolphi spoke of the ligulæ of fishes which could continue to live in birds.

At a period nearer to our own times, our learned friend, Von Siebold, deservedly called the prince of helminthologists, was entirely of this opinion, and compared the cysticercus of the mouse with the tape-worm of the cat, considering this young worm as a wandering, sick, and dropsical being.

In his opinion, the worm had lost its way in the mouse, as the tænia of the cat could live only in the cat. Flourens considered it a romance when I myself an-

nounced to the "Institut de France," that cestode worms
must *necessarily* pass from one animal to another in
order to complete the phases of their evolution.

At the present time, experiments respecting these
transmigrations are repeated every day in the laboratories of zoology with the same success; and Mons. R.
Leuckart, who directs with so much talent the Institute
of Leipzig, has discovered, in concert with his pupil
Mecznikow, transmigrations of worms accompanied by
changes of sex; that is to say, they have seen nematodes,
the parasites of the lungs of the frog, always female or
hermaphrodite, produce individuals of the two sexes
which do not resemble their mother, and whose habitual
abode is not in the lungs of the frog but in damp earth.
In other words, let us imagine a mother, born a widow,
who cannot exist without the assistance of others, producing boys and girls able to provide for themselves.
The mother is parasitical and viviparous, her daughters
are, during their whole life, free and oviparous.

This observation leads us to another sexual singularity, lately observed, of males and females of different
kinds in one and the same species, and which give birth
to progeny which do not resemble each other; the same
animals, or rather the same species, proceed from two
different eggs fecundated by different spermatozoids.

Now that these transmigrations are perfectly known
and admitted, the starting-point of the inquiry has been
so entirely forgotten that the honour of the discovery
has been frequently attributed to fellow-workers, who
had no knowledge of it till the demonstration had been
completed, and the new interpretation generally accepted.
But let us return to our subject.

2

The assistance rendered by animals to each other is as varied as that which is found amongst men. Some receive merely an abode, others nourishment, others again food and shelter; we find a perfect system of board and lodging combined with philozoic institutions arranged in the most perfect manner. But if we see by the side of these paupers, some which render to one another mutual services, it would be but little flattering to them to call all indiscriminately either parasites or messmates (*commensaux*). We think that we should be more just to them if we designated the latter kinds *mutualists*, and thus *mutuality* will take its place by the side of *messtable* arrangements (*commensalism*) and of parasitism.

It would also be necessary to coin another name for those which, like certain crustaceans, or even some birds, are rather guests which *smell out a feast from afar* (pique-assiettes) than parasites; and for others which repay by an ill turn the assistance which they have received. And what name shall we give to those which, like the plover, render services which may be compared to medical attendance?

This bird in fact performs the office of dentist to the crocodile. A small species of toad acts as an accoucheur to his female companion, making use of his fingers as a forceps to bring the eggs into the world. Again, the pique-bœuf performs a surgical operation, each time that he opens with his lancet the tumour which encloses a larva in the midst of the buffalo's back. Nearer home, we see the starling render in our own meadows the same service as the pique-bœuf (*Buphaga*) in Africa; and we may see that among these living creatures there is more than one speciality in the healing art.

We must not forget that the occupation of a grave-digger is equally general in nature, and that it is never without some profit to himself or his progeny that this gloomy workman inters the bodies of the dead. Certain animals have an occupation analogous to that of the shoeblack or the scourer, and they freshen up with care, and even with a kind of coquettish pleasure, the toilet of their neighbours.

And how must we designate the birds known by the name of stercorariæ, which take advantage of the cowardice of sea-gulls in order to live in idleness? It is useless for the gulls to trust to the strength of their wings, the stercorariæ in the end compel them to disgorge their food in order that they may partake of the spoils of their fishery. When followed up too closely, these timid birds throw up the contents of their crop, to render themselves lighter, like the smuggler who finds no means of safety except in abandoning his load.

We must not, however, be too hard upon all this class, since very often, as in the case of the gnat, it is only one of the sexes which seeks a victim.

All animals usually live for the passing day; and yet there are some which practise economy, which are not ignorant of the advantages of the savings bank, and, like the raven and the magpie, think of the morrow, to lay up in store the superfluity of the day's provision.

As we have before said, this little world is not always easy to be known, and in its societies, to which each brings his capital, some in activity, others in violence or in stratagem, we find more than one *Robert Macaire* who contributes nothing, and takes advantage of all. Every species of animal may have its parasites and its messmates, and each may perhaps have some of different sorts, and in diverse categories.

But whence come those disgusting beings, whose name alone inspires us with horror, and which instal themselves without ceremony, not in our dwellings, but in our organs, and which we find it more difficult to expel than rats or mice ? They all derive their existence from their parents.

The time has passed when a vitiated condition of the humours, or the deterioration of the parenchyma was considered a sufficient cause for the formation of parasites, and when their presence was regarded as an extraordinary phenomenon resulting from the morbid dispositions of the organism. We have reason to hope that this language will, during the next generation, have entirely disappeared from works on physiology and pathology. Neither the temperament nor the humours have any influence on parasites, and they are not more abundant in delicate individuals than in those who enjoy the most robust health. On the contrary, all wild animals harbour their parasitical worms, and the greater part of them have not lived long in captivity, before nematode and cestode worms completely disappear. It is only the imprisoned parasites which do not desert them.

All these mutual adaptations are pre-arranged, and as far as we are concerned, we cannot divest ourselves of the idea that the earth has been prepared successively for plants, animals, and man. When God first elaborated matter, He had evidently that being in view who was intended at some future day to raise his thoughts to Him, and do Him homage.

This is the answer which I would give to the question recently propounded by Mons. L. Agassiz. "Were the physical changes to which our globe has been subjected effected for the sake of the animal world, considered in its relations from the very beginning, or are the modifications of animals the result of physical changes? in other words, has the earth been made and prepared for living beings, or have living beings been as highly developed as was possible, according to the physical vicissitudes of the planet which they inhabit?

This question has always been discussed, and that science which cannot look beyond its scalpel, will never succeed in resolving it. Each one must seek by his own reason the solution of the great problem.

When we see the newly-born colt eagerly seeking for its mother's teats, the chick as soon as it is hatched beginning to peck, or the duckling seeking its puddle of water, can we recognize anything but instinct as the cause of these actions, and is not this instinct the libretto written by Him who has forgotten nothing?.

The statuary who tempers the clay from which to make his model, has already conceived in his mind the statue which he is about to produce. Thus it is with the Supreme Artist. His plan for all eternity is present to His thought. He will execute the work in one day, or in a thousand ages. Time is nothing to Him; the work is conceived, it is created, and each of its parts is only the realization of the creative thought, and its predetermined development in time and space.

"The more we advance in the study of nature," says Oswald Heer in "Le Monde primitif" which he has just published, "the more profound also is our conviction, that belief in an Almighty Creator and a Divine Wisdom, who has created the heavens and the earth according to an eternal and preconceived plan, can alone resolve the enigmas of nature, as well as those of human life. Let us still erect statues to men who have been useful to their fellow-creatures, and have distinguished themselves by their genius, but let us not forget what we owe to Him who has placed marvels in each grain of sand, a world in every drop of water."

At first we shall treat of animal *messmates*, secondly of *mutualists*, and thirdly of parasites.

ANIMAL PARASITES

AND MESSMATES.

CHAPTER I.

ANIMAL MESSMATES.

THE messmate is he who is received at the table of his neighbour to partake with him of the produce of his day's fishing; it would be necessary to coin a name to designate him who only requires from his neighbour a simple place on board his vessel, and does not ask to partake of his provisions.

The messmate does not live at the expense of his host; all that he desires is a home or his friend's superfluities. The parasite instals himself either temporarily or definitively in the house of his neighbour; either with his consent or by force, he demands from him his living, and very often his lodging.

But the precise limit at which commensalism begins is not always easily to be ascertained. There are animals which live as messmates with others only at a certain period of their lives, and which provide for their own support at other times; others are only messmates

under certain given circumstances, and do not usually
merit this appellation.

In the higher animals, this relation between them is
generally well known, and justly appreciated, but it is not
the same in the inferior ranks ; and more than one
animal may pass for a messmate or a parasite, for a
robber or for a mendicant, according to the circumstances under which he is observed. The sharper passes
for an honest man as long as he has not been taken *in
flagrante delicto.* Thus, in order to be just, we must
carefully examine the indictment, and not pronounce
sentence without strict examination.

The greater part of those animals which have established themselves on each other, and live together on
a good understanding and without injury, are wrongly
classed as parasites by the generality of naturalists.
Now that the mutual relations of many of these are
better understood, we know many animals which unite
together to render each other mutual assistance ; while
there are others which live like paupers on the crumbs
which fall from the rich man's table. There are many
relations between the different species which can be discovered only after minute examination, but which have
recently been appreciated with greater impartiality.

Animal messmates are rather numerous, and commensalism has been observed, not only in animals of the
present age, but in those of the primary epoch. Wyville
Thomson explained to me, while I was myself his messmate at Edinburgh, at the meeting of the British Association in 1871, that the polyps of the Silurian age
already practised it. We do not class among animal
messmates those living creatures which, like the birds

which we keep in cages, charm the ear with their song, or which, in spite of our care, live at the expense of our pantry; we will only refer to veritable messmates, which, sometimes through weakness of constitution, sometimes for want of activity, can neither feed themselves nor bring up their family without seeking help from their neighbours.

There are some free messmates which never renounce their independence, whatever may be the advantages which their Amphitryon enjoys; they break their alliance with him for the slightest motive of discontent, and go and seek their fortune elsewhere. Their susceptibility or their love of change guides them. They are recognized by their fishing implements or their travelling gear, which they never lay aside. These free messmates are the more numerous. The others, the fixed messmates, instal themselves with a neighbour, and live at their ease, having completely changed their dress, and renounced for ever an independent life. Their fate is thenceforward bound to him who carries them.

Under these two categories we shall cite several examples, and glance at the differences which the various classes of the animal kingdom present in this respect, beginning with the higher ranks.

CHAPTER II.

FREE MESSMATES.

WE meet with free messmates in various classes of the animal kingdom. They sometimes mount on the back of a neighbour, sometimes occupy the opening of the mouth, the digestive passages, or the exit for the excreta; at times they place themselves under the shelter of the cloak of their host, from whom they receive both aid and protection.

Among the vertebrates, there are few except fishes which merit a place here; it is only amongst these that we meet with species at the mercy of others, and dependent on acolytes, which are in every respect inferior to themselves.

An interesting messmate belonging to this first category is a fish of graceful form, named donzelina, which goes to seek its fortune in the body of a holo-thuria. Naturalists have long known it under the name of Fierasfer. It has a long body like that of an eel, entirely covered with small scales; and as it is quite compressed, it has been compared to the sword which conjurors thrust into their œsophagus. They are found in different seas, and all have similar habits. This fish is lodged in the digestive tube of his companion, and, without any regard for the hospitality which he receives, he seizes on his portion of all that enters. The Fierasfer contrives to cause himself to be served by a neighbour better provided than himself with the means of fishing.

Dr. Greef, at present Professor at Marbourg, found at Madeira a holothuria of a foot in length, in which a vigorous Fierasfer lived in peace. Quoy and Gaimard, in the account of their voyage round the world, have remarked long since, that the *Fierasfer hornei* is found in the *Stichopus tuberculosus*.

The holothuriæ seem to exist under very advantageous conditions in this respect, since we see Fierasfers, which are themselves tolerable gluttons, accompanied by Palæmons and Pinnotheres in the same animal. Professor C. Semper has seen holothuriæ in the Philippine Islands which bore a considerable resemblance, in this respect, to an hotel with its table d'hôte.

These singular fishes have been long noticed, but it was not till recently that their presence in a host so low in the scale as a holothurian could be explained.

But if naturalists are agreed as to the bond which unites these fishes to the holothuriæ, they do not agree as to the organs which they inhabit in their living hotel. Do they lodge in the digestive cavity of the holothuriæ, or do they inhabit the arborescent respiratory processes which open at the posterior extremity of the body ? Until recently it was thought that it was in their stomach, but a doubt has arisen.. Professor Semper, who has studied these animals with particular care at the Philippine Islands, had the curiosity to open the stomach of some of them, and found there, not the animals taken by the holothuriæ, but the remains of its respiratory processess which they were in the act of digesting. Is it then merely a messmate? We must have more information on this point; and if it were not accidentally that the fierasfer swallowed the walls of the compartment in which he was lodged, he ought rather to take his place among parasites. Though it lodges in the respiratory processes, as the learned professor at Wurtzburg asserts, the fierasfer may also be a messmate after the fashion of so many others which inhabit the neighbourhood of the rectum, in order the more conveniently to snap up those animals which are attracted by the odour.

The fierasfers are not the only fishes which seek assistance from the holothuriæ; a species lives at Zamboanga, to which the specific name of Scabra has been given, and in the stomach of which, says Mons. Johannes Müller, usually lives a myxinoid fish, called *Enchelyophis vermicularis*. Unfortunately, we are not told in what part of the stomach it resides; for all is stomach in these animals.

It is less degrading for a fish to ask assistance from one in his own rank. The Mediterranean offers a curious instance of this. Risso saw at Nice, at the commencement of this century, the monstrous fish known under the name of *Beaudroie* (the angler, or fishing-frog) lodging in its enormous branchial sac a fish of the family of the Murenidæ, the *Apterychtus ocellatus*. He is found there evidently under the condition of a messmate. Although the eels generally get their living easily, the Angler possesses fishing implements which are wanting in them, and

when both of them are immersed in the ooze, it carries
on a fishery sufficiently abundant to enable it to share
the spoil with others. This same angler lives in the
northern seas, and there it harbours an amphipod crustacean, which until lately has escaped the vigilance of
carcinologists. We shall speak of it further on.

Dr. Collingwood saw a sea anemone in the Chinese
Sea, which was not less than two feet in diameter, and
in the interior of which lodges a very frisky little fish,
the name of which he could not tell.

Lieut. de Crispigny has observed a sea anemone
(*Actinia crassicornis*) living on good terms with a
malacopterygian fish, the *Premnas biaculeatus*. This
fish penetrates into the interior of the anemone ; the
tentacles close round it, and it lives thus for a considerable time enclosed as in a living tomb. Mons. de
Crispigny has kept these animals alive for more than a
year, in order to make careful observations on them. A
fish known by the name of *Oxybeles lumbricoides* has been
also found in the Indian Seas, which modestly takes up
his quarters in a star-fish (*Asterias discoida*). Another
case of *commensalism* has been made known to us by
Professor Reinhardt of Copenhagen. A siluroid of Brazil,
of the genus *Platystoma*, a skilful fisherman, thanks to
his numerous barbules, lodges in the cavity of his mouth
some very small fishes, which were for a long time considered as young siluroids; it was supposed that the
mother brought her progeny to maturity in the cavity
of the mouth, as marsupials do in the abdominal pouch,
or as some other fishes do. These messmates are perfectly developed and adult, but instead of living on the
produce of their own labour, they prefer to instal them-

selves in the mouth of an obliging neighbour, and to take their tithes of the succulent morsels which he swallows. This little fish has received the name of *Stegophilus insidiatus*. We see that in the animal world it is not always the great which take advantage of the little. Still, let us not be deceived; there are fishes in the latitude of the Island of Ceylon which really hatch their eggs in the cavity of the mouth, and we have seen some in the museum at Edinburgh, labelled with the name of *Arius bookei*. Louis Agassiz has made the same observation on a fish of the Amazon, which has also been recognised by Jeffreys Wyman. One fish wraps up its eggs in the fringes of its branchiæ, and protects them till they are hatched; another lays its eggs in holes hollowed out by itself in the steep banks of the river, and protects the young ones after they are hatched.

To hatch the eggs in the mouth is not more extraordinary than to hatch them in any other part of the body. The *Sygnathidæ* hatch theirs in a pouch behind the anus; and it is a curious circumstance that the females do not undertake this duty. The males alone carry their progeny with them. This recalls to our recollection that curious example of the birds known under the name of *Phalaropes*, among which the males only hatch the eggs.‧ The female of the cuckoo abandons her eggs, and entrusts them to the female of another bird.

The cuckoo suggests to us the mound-making Megapode and the Talegalla of Latham, both of which inhabit Australia; these birds deposit their eggs in an enormous mass of leaves or grass, which grows warm by decomposition, and the temperature of which is great enough to hatch them. The young ones when they come out of the egg are sufficiently developed to be able to provide for their own wants, and to do without a mother's care.

To return to our animal messmates: let us notice the result of the observations of a learned and skilful naturalist who has rendered great services to ichthyology. Dr. Bleeker has described a still more remarkable association in the Indian seas; it is that of a crustacean, the *Cymothoa*, taking advantage of a fish known under the name of *Stromatea;* too imperfectly organized to fish for itself at large, but more skilful in snapping up all that comes within its reach, it makes its home in the buccal cavity of the Stromatea.

But of all crustaceans, the most cruel is the isopod named *Ichthyoxena*, which hollows out for itself and its female a large dwelling-place in the coats of the stomach of a cyprinoid fish. We will return again to these examples.

The *Physaliæ*, those charming living nosegays of the tropical regions, also give lodging in their cavities, and in the midst of their long cirrhi, to little adult and perfect fishes, belonging to the family of the *Scombridæ*, a family to which are attached the tunny and the mackerel. These sea-butterflies flutter away their indolent existence at the expense of their host. Voyagers tell us that they have seen them by dozens concealed in these animated festoons. Mons. Al. Agassiz has mentioned, in his illustrated catalogue, another fact, quite as extraordinary, observed in the Bay of Nantucket, in the United States ; it relates to a nocturnal Pelagia (*Dactylometra quinque-cirra*, Ag.) always accompanied, not to say escorted, by a species of herring. The two neighbours constitute together an association which probably redounds to the advantage of both.

Without quitting our own sea-coast, we find an association of the same kind between young fishes (*Caranx trachurus*) and a beautiful medusa (*Chrysaora isocela*). This sea nettle often encloses several young specimens of Caranx, which we are surprised to see issuing full of life from the transparent bodies of these polyps. Indeed, it is not rare to find other fishes in the medusæ. Dr. Gunther, who has arranged with so much care the rich collection of fishes in the British Museum, has shown us some specimens of the *Labrax lupus*, and of the *Gasterosteus*, which had been obtained from the interior of different medusæ; and these associations have been also remarked by various distinguished observers, among whom we may mention Messrs. Sars, Rud. Leuckart, and Peach. The captain of the frigate *Jouan*, when in the Indian Sea, on October 26th, 1871, in 13° 20′ N. lat., and 60° 30′ E. long., that is to say, about 200 leagues to the west of the Laccadive Islands, saw, in very fine weather, the sea, which was at that time very calm, covered with medusæ, and the greater part of these were escorted by many little fishes of the genus *Ostracion*, the species of which he was unable to ascertain. It is probable that the school of medusæ set in motion certain animals which are eagerly sought after by the Ostracions.

The Pilot is a fish of which much has been recorded; fishing for it is one of the principal recreations of sailors during their long voyages. Some assure us that it snaps off the bait, without touching the murderous hook which threatens the shark; and as it never quits its companion, others have supposed that it lives on the morsels abandoned by it. Neither of these suppositions is correct; and as the shark does not need its services to point out the danger, we must content ourselves with mentioning this curious association without endeavouring to explain it.

In fact, we have had the opportunity of examining many well-preserved specimens, the stomach of which contained potato parings, the carapaces of crustaceans, the *débris* of fishes, marine plants (fuci), and a piece of *cut* fish, which had evidently served as a bait. The pilot does not, therefore, live on the leavings of his companion, but on his own industry, and doubtless finds some advantage in piloting his neighbour. Through the great kindness of Dr. Gunther we have been able to make this interesting examination in the rich galleries of the British Museum. We desire to take this opportunity of expressing our gratitude to this learned man and to his illustrious colleagues, who have the direction of that vast establishment, which is ever open to those who labour for the advancement of science.

The pilot has sometimes been confounded with a very different fish, which does not merely remain in the neighbourhood of the shark, but establishes itself upon him, and moors himself to him by the aid of a particular apparatus, for a longer or shorter time; we may even say during the whole of the voyage. This is the Remora.

Is this fish the messmate of the shark to which he is attached? As in the case of the pilot, an examination alone could decide the question. We have opened at the British Museum the stomachs of several remoras of different sizes, and we have been able to ascertain that they also fish on their own account; their food was composed of morsels of fish which had served as bait, of young fish swallowed whole, and of some remains of crustacea. The remora is simply anchored to his host, and asks from him nothing but his passage. He is contented, like the pilot, to fish in the same waters as the shark which transports him. Sailors, even now, are convinced that if any one of these remoras should attach itself to the ship, no human power could cause it to advance, and that it must of necessity stop. It is certain that the fishermen of the Mozambique Channel take advantage of this faculty, to fish for turtles and certain large fish. They pass through the tail of the remora a ring to which a cord is attached, and then send it in pursuit of the first passer-by which they consider worthy to be caught. This kind of fishing resembles in some degree the sport of hawking with falcons.

So extraordinary a being could not fail to attract the attention of those among the ancients who were students of nature. Pliny assures us that the remora was used in the preparation of a philtre capable of extinguishing the flames of love.

There must be many free animal messmates among insects, and entomologists should make them known; for example, many of them live with ants, as the *Psela-phidæ* and *Staphylinidæ.* Certain hairs of these insects, it is said, secrete a sweet liquid of which ants partake greedily. If we may believe a skilful observer, Mons. Lespès, there are some among them, as the Clavigers, which in exchange for the services which they render are fed by the ants themselves. We may also mention the larvæ of the *Meloë,* which seem to live as parasites, and the true nature of which was so long unknown,

The females of the *Meloë* lay their eggs near the ranunculus and other plants whose flowers are regularly visited by bees. After these are hatched, the larvæ ascend into the flowers and wait patiently till a bee takes them on his back, and carries them into the interior of the hive. This insect was formerly known under the name of the bee-louse, but this appellation is improper, for the bee is not the host of the meloë, but simply its beast of burden. According to recent observations, flies perform the same office for *Chelifers*, and certain aquatic and land coleoptera for several kinds of acaridæ.

In the class of animal messmates we find also a coleopterous insect that lodges in a manner similar to the paguri, of which we shall presently speak. The female of the *Drilus*, a species allied to glowworms, attacks the snail, and when it has devoured it, instals itself in the shell, to pass through its metamorphoses; when necessary, it frequently changes its shell and chooses successively more spacious lodgings. Like a true Sybarite, the drilus weaves a curtain of tapestry before the entrance of its habitation, and remains there peaceably surrounded by the vestment of its youth.

Remarkable examples of free messmates are found more especially among crustaceans. It is well known that this class includes lobsters, crabs, prawns, and those legions of small animals which serve as the police of the sea-shore, purifying the waters of the ocean of all organic matters, which otherwise would corrupt them. They do not, like insects, shine with variegated colours; their forms are hardy and varied, and they are often pleasing on account of the singularity of their movements. Professor Verrill has recently studied some of these creatures, and has clearly shown how interesting they are, not only to naturalists, but to people in general.

Crustaceans and worms furnish the greatest number of paupers and infirm individuals; and a great many of them need the continual assistance of their neighbours to enable them to get their living. While other animals advance towards perfection as they grow older, it is far different with many crustaceans, and we should be tempted to refer to the vegetable kingdom many of them at the very period when they are approaching the adult condition. Cuvier placed all the class of cirrhipedes among the mollusca, and the lernæans among the worms. Many of these animals which are but indifferently adapted to live without help from others, have recourse to benevolent neighbours; from one they seek only shelter, from another a part of his booty, from a third both an asylum and protection. They are often reduced to a mere skin; everything else has disappeared, and there remains no proper organ except that which is necessary for the reproduction of the species. Corpulent, blind, impotent, legless cripples, their existence is more precarious than that of those miserable mutilated beings found in our cities; they only live on the blood of the neighbour which gives them an asylum. Yet when they first quit the egg they are all free; they frisk, they swim with the rapidity of lightning, and at the close of life we find them deformed, and crouched in some living refuge, as if a foul leprosy had atrophied within them all the organs which served as a means of communication with the outer world. Parasites and messmates, furnished at first with the same kind of limbs and the same habits, can sometimes only be distinguished from each other when we have made our observations on them in their first swaddling clothes. The child has given a clue to the history of the old man.

We will not examine these animals in all the details of their private life, and yet we are strongly tempted to confess to our readers some of the indiscreet acts of which we have been guilty, in watching them while changing their dress. Notwithstanding their shyness and their desire to escape observation during the moulting period, we have more than once made observations on them while quitting their garment which has become too small. The old tunic generally splits down the back, and falls off all in one piece as it gives the animal egress. The crustacean is extended quite soft and supple by the side of its rigid carapace.

Of all the free crustacean messmates, one of the most interesting, though among the smallest of them, is a tiny crab, about as large as a young spider, which lives in mussels, and which has been often accused, though evidently wrongfully, as the cause of the indisposition so well known by those who are fond of this mollusc. Very many of them have been seen within the last few years, and yet accidents have been very few. The mussels themselves are guilty; they produce on some persons an injurious effect, through *idiosyncracy*. We have at least a word to serve as an explanation, and at present we must content ourselves with it.

Under what conditions do those crabs, called by naturalists Pinnotheres, and which we do not find elsewhere, inhabit mussels? Are they parasites, pseudoparasites, or messmates? It is not a taste for voyaging which tempts them, but the desire of having always a secure retreat in every place. The pinnothere is a brigand who causes himself to be followed by the cavern which he inhabits, and which opens only at a well-known watchword. The association redounds to the advantage of both; the remains of food which the pinnothere abandons are seized upon by the mollusc. It is the rich man who instals himself in the dwelling of the poor, and causes him to participate in all the advantages of his position. The pinnotheres are, in our opinion, true messmates. They take their food in the same waters as their fellow-lodger, and the crumbs of the rapacious crabs are doubtless not lost in the mouth of the peaceful mussel. There is no doubt that these little plunderers are good lodgers, and if the mussels furnish them with an excellent hiding-place and a safe lodging, they themselves profit largely by the leavings of the feast which fall from their pincers. Little as they are, these crabs are well furnished with tackle, and advantageously placed to carry on their fishery in every season. Concealed in the bottom of their living dwelling-place (a den which the mussel transports at will) they choose admirably the moment and the place to rush out to the attack, and always fall on their enemy unawares. Some of these pinnotheres live in all seas, and inhabit a great number of bivalve molluscs. The northern seas contain a large species of Modiola (*Modiola Papuana*) which is especially found in deep and almost inaccessible parts, and which always encloses a couple of pinnotheres about the size of a hazel-nut. We have opened hundreds of these modiolæ, and we have never met with any without their crabs. We have long since deposited some specimens of these pinnotheres in the galleries of the Natural History Museum at Paris.

The large mussel, which furnishes fine pearls (*Avicula margaritifera*), lodges also pinnotheres of a particular species by the side of another messmate more allied to a lobster than a crab. It is not even impossible that these crustaceans, with other messmates or parasites, contribute to the formation of pearls, since these gems, so highly prized in the fashionable world, are only the result of vitiated secretions, and are usually the result of wounds.

We also meet with a little crab (*Ostracotheres tridacnæ*, Ruppel) in the acephalous mollusc, whose immense shell sometimes serves as a vessel for holy water; and it lives doubtless in many other bivalves which have not yet been examined.

Dr. Léon Vaillant has written a very interesting memoir on the Tridacnæ, and informs us that the crab takes shelter in their branchial chamber. Therefore, since the molluscs live only on vegetable substances, while the Ostracotheres feed entirely on animal matter, Mons. Vaillant supposes that the latter take their choice of the food as it enters, and seize on its passage that which suits them best. Mr. Peters, during his abode on the coast of Mozambique, studied a great many of these acephala and pearl-mussels, and found their interior inhabited by three crustacean decapods, a pinnothere, and two macrouræ allied to the *Pontonia*, to which he has given the name of *Conchodytes;* the *Conchodytes tridacnæ* inhabits the *Tridacna squamosa;* the *Conchodytes meleagrinæ*, as its specific name indicates, lives in the shell of the pearl-mussel.

Professor Semper has recently observed pinnotheres in holothurians at the Philippine Isles, and Mons. Alphonse M. Edwards has described some from New Caledonia (*P. Fischerii*) ; so that these little crabs, the friends of the molluscs, are known in both hemispheres.

Do not these conditions seem to authorize the conclusion that the same thought has presided over the appearance of all living creatures ; that they have all come into existence, not according to the chance arrangement of surrounding media, but according to the laws established from the very origin of all things ?

The shell which lodges both these pinnotheres, in the Mediterranean as well as the Atlantic, is a large acephalous mollusc, known under the name of *Jambonneau* (a small ham or gammon), and which, according to Aristotle, harbours two different kinds of messmates. This illustrious natural philosopher also described a Pontonia (*Pontonia custos*, Guérin—*P. Pyrrhena*, M. Edw.) about an inch and a half long, of a pale rose colour, more or less transparent, and which lives with its companion, the pinnothere, in the cavity of the *Pinna marina*. This is the same animal which a naturalist of the last century named the *Cancer custos*.

We have wished to ascertain whether Pliny knew these crustaceans. He has spoken of them in the following terms :—" The Chama is a clumsy animal without eyes, which opens its valves and attracts other fishes, which enter without mistrust, and begin to take their pastime in their new abode. The pinnothere seeing his dwelling invaded by strangers, pinches his host, who immediately closes his valves, and kills one after another these presumptuous visitors, that he may eat them at his leisure."

Cuvier did not believe that the pinnothere brought any food to the mollusc, since the latter, in his opinion, lives entirely on sea-water.

Other zoologists regard the pinnothere as an intruder whom chance has brought into this mysterious position. Others again consider mussels as acquaintances possessed of a very curious disposition, and that having no eyes, they have interested in their fate this little crab, which is perfectly provided with eyesight. In fact, in common with other crustaceans of his species, he carries on each side of his carapace, at the end of a movable stalk, a charming little globe, provided with some hundreds of eyes, which he can direct upon his prey, as the astronomer turns his telescope on any point of the firmament. These later naturalists consider, in fact, their crab as a living journal which supplies his host with the news of the day. Rumphius, a Dutchman, the first who described the animal of the nautilus, also understood the habits of pinnotheres. In his "Amboinche Rariteit Kamer," published in 1741, he says that these crustaceans inhabit always two kinds of shellfish, the *Pinna* and the *Chama squamata*. According to him, when these molluscs have attained their growth, one pinnothere (one only at least in the Chama) lives in their interior and does not abandon its lodging till the death of its host. Rumphius regards this crustacean as a faithful guardian, fulfilling the duties of a door-keeper. In 1638 he found actually two sorts of keepers : by the side of a Brachyuron, carrying an embossed buckler, slender in front, he discovered a Macrouron of the length of his finger-nail, of a yellowish orange colour, semitransparent, with white and very slender claws. It is

3

without doubt the same animal that Mons. Peters, of Berlin, found on the coast of Mozambique, and of which we have spoken before.

A little crab is known to live near the coast of Peru (*Fabia Chilensis*, Dana), which exists under somewhat different conditions. He chooses, not a bivalve mollusc, but a sea-urchin (*Euriechinus imbecillus*, Verrill), and lodges in the intestine, near its termination, so as to seize as they pass by all those living creatures which are attracted by the odour. Doubtless, the delicacy of our sense of smell is disgusted by such a mode of seeking food; but this predilection may have a reason with which we are not acquainted. There are a considerable number of other species which live under similar conditions.

On the coast of Brazil, my son found two couples of crabs in the tube of a very long annelid, narrow at the ends, and wide in the middle. The tube was too small at the end to allow them to escape. These crustaceans had, no doubt, penetrated thither before they had attained their full size.

A crab of the family of the Maidæ conceals itself in the substance of a polypidom very common in the Viti Islands, in company with a gasteropod mollusc, and both of them assume the exact colour of the polypidom. This is a new kind of *mimicry*. This crab is known by the name of *Pisa Styx*, the gasteropod is a *Cypræa*, the polyp is the *Melithea ochracea*. A decapod crustacean, the *Galathea spinirostris*, seeks for a *Comatula*, the colour of which it exactly imitates, and with which it lives on the most friendly terms.

The holothuriæ, of which we have already spoken, appear to afford an abode to many animals: indepen-

dently of the *Fierasfer*, the *Holothuria scabra* of the
Philippine Islands regularly lodges in its interior a
couple, and sometimes, though rarely, a greater number
of pinnotheres belonging to two distinct species. They
choose this domicile at an early period, and must be highly
delighted with this obscure abode, since they are seen
no more, and when they have once entered never quit
this living cavern. This observation is due to Professor
Semper, who has made us acquainted with so many
curious facts of the China Sea and the Pacific Ocean.
In the midst of the slender branches of a coral of the
Sandwich Islands, the *Pæcilopora cæspitosa* of Dana,
there lives a little crab (*Hopalocarcinus marsupialis*,
Stimpson), which is at last completely enclosed by the
vegetation of the coral. It only keeps up sufficient
communication with the exterior to enable it to procure
food. The coral, however, furnishes it nothing but a
resting-place in the midst of its tissues.

Among the Philippine Islands, also, a brachyurous
crustacean lives in the branchial cavity of one of the
Haliotidæ, and another on the body of a holothuria. On
the coasts of Brazil, F. Müller, during his abode at
Desterro, saw some *Porcellanæ* inhabiting star-fish, not
as parasites, as had been supposed, but as true messmates. A crustacean possessed of but little generosity
is the *Lithoscaptus* of Mons. Milne-Edwards. Provided
with beak and claws for the purpose of attack, it instals
itself, sad to say, in the pantry of a medusa, and instead
of making use of its own weapons, takes advantage of
the perfidious nematocysts of its acolyte, in order to live
quietly at his expense.   .

Under the name of *Asellus medusæ*, Sir J. G. Dalyell

has made us acquainted with another messmate of the medusæ which greatly resembles an *Idothea*.

Another kind of commensalism is that of the Dromiæ. These crabs are of the ordinary size, and lodge, from their earliest youth, under a growing family of polyps, which increases with them. This colony has for its principal foundation a living Alcyonium, which covers the carapace, and as it develops, adapts itself perfectly to all the inequalities of the cephalothorax; one might consider it an integral part of the crab. Sertulariæ, Corynes, Algæ, develop themselves on this Alcyonium, and the Dromia, masked by this living rock which it carries on its shoulders like the fabled Atlas, marches gravely in pursuit of her prey. She has no fear of arousing the attention of her enemies. The greatest vigilance cannot prevent the sudden attack of these dangerous neighbours. There is in the Mediterranean a species which sometimes comes to our coast. They are also known in the Indian Seas and in the Northern Pacific. Rumphius named the dromia *Cancer lanosus;* it is, said he, a crab which carries grass or moss on its back. It is also mentioned by Renard. Dana has observed a sea-anemone covering a crab in the same manner as the Alcyonium does the dromia, and which is not less dangerous. The mode of life of this anemone has procured for it the name of *Cancrisocia expansa.* In the north of California, a crab (*Cryptolithoides typicus*) covers itself in the same manner with a living cloak which hides it from view, and under cover of which it surprises those whom it attacks. It has already cleared the ground of its prey before any alarm has been given to the neighbourhood.

We should perhaps speak here of an association of another kind, the nature of which it is difficult to ascer-certain; I refer to the little crab, the Turtle Crab of Brown, which is met with in the open sea on the cara-pace of turtles, and sometimes on sea-weeds. It may be supposed that it takes advantage of the carapace of its neighbour, in order to transport itself at little expense into different latitudes, and it is asserted that the sight of this crustacean gave confidence to Christopher Colum-bus, eighteen days before the discovery of the New World. Besides this animal, a whole society chooses this movable habitation: in addition to the cirrhipedes we also find the *Tanaïs*, which is not, however, con-demned to live there always.

The macrourous decapods are more rarely found as messmates, but still a Palæmon is sometimes seen on the body of an Actinia, according to Semper, and another in the branchial cavity of a Pagurus. But that which is more generally known, is the presence in the *Euplec-tella aspergillum* of the palæmon which lodges in this fairy palace. It is probable that the Euplectella of the Atlantic, recently observed near the Cape Verd Islands by the naturalists on board the *Challenger*, also conceals this crustacean in its interior. We may also allude here to the *Hypoconcha tabulosa*, a crab whose carapace is too soft to allow it to venture out undefended, and which covers itself with the shell of a bivalve mollusc.

Among the various associations of this kind, nóne is more remarkable than that of the soldier-crabs, so abun-dant on our coasts, and called by the names of *Bernard the Hermit* and *Kakerlot* by the Ostend fishermen. It is well known that these crabs are decapod crustaceans, very like miniature lobsters, which lodge in deserted shells, and change their dwelling-place as they grow larger. The young ones are content with very little habitations.

The shells which give them shelter are such as have been shed, which they find at the bottom of the sea, and and in which they conceal their weakness and their misery. These animals have an abdomen too soft to bear the dangers which they meet with in their warfare, and that they may be less exposed to the claws of their numerous enemies, they take shelter in a shell which serves at the same time both as a dwelling and a buckler. Armed cap-à-pie, the soldier-crabs march boldly on the the enemy, and know no danger, since they always have a secure retreat.

But this animal does not live alone in this asylum. He is not so much of an anchorite as he appears to be, for by his side an annelid usually instals himself as a messmate, which forms with the Pagurus one of the most terrible associations that are known. This annelid is a long worm, like all the nereids, whose supple and undulating body is armed along its sides with arrows, lances, pikes, and poniards, the wounds of which are always dangerous. It is a living panoply which glides furtively into the enemy's camp without giving the alarm.

When a pagurus is on the march it resembles a nest of pirates, who never cease their exploits till all has been ravaged around them. This shell is so innocent in its appearance, that it introduces itself everywhere without provoking the least suspicion. It is usually covered with a colony of Hydractiniæ, and in the interior, Peltogasters,

Lyriopes, and other crustaceans often establish themselves. The paguri are not messmates of an ordinary
kind, for they inhabit only a deserted shell. They are
spread over all seas. They are found in the Mediterranean, the Northern Sea, on the coasts of the Pacific, of
New Zealand, and of the East Indian islands: thirty
species and even more have been inserted in the catalogue
of crustaceans.

Naturalists have given the name of *Cenobitæ* to some
pagurians inhabiting the seas of warmer latitudes; these
have an abdomen like the pagurus, antennæ like the
Birgus, and like it they inhabit shells. The *Cenobita
Diogenes* is a species found in the Antilles.

Other pagurians, the *Birgi*, grow very large, and conceal their abdomen no longer in a shell, but in the
crevices of the rocks, as lobsters do at the moulting time,
to protect their body while deprived of their defensive
armour. In the East Indies they remain on land, and
even climb into trees. They have so much strength in
their pincers, that Rumphius relates of one of these
crustaceans, that, while stretched on a branch of a tree,
it raised a goat by the ears.

Side by side with the pagurians which instal themselves in a shell with thick and completely opaque walls,
we recognize crustaceans of the order of amphipods,
the *Phronimæ*, which choose for themselves not an abandoned hovel, but a veritable crystal palace, and take
possession of it without inquiring whether or no it is
inhabited. The daylight penetrates through the walls of
their dwellings, and it can scarcely be discerned in
the water whether or no their body is protected by a
covering. They usually take the dwelling of a Salpa, a

Beroë, or a Pyrosoma, and from within this lodging they give themselves up to the pleasures of fishing.

The *Phronima sedentaria* which lodges with the salpa seems to be scattered over the warm seas of both hemispheres. For the honour of the species, the females alone seek the assistance of their neighbours, without at the same time abandoning their characteristic robe. The sexes differ little from each other except in size, in the abdomen, and in the antennæ. Maury has described certain amphipod crustaceans which also inhabit the Salpæ.

Another phronima described by Professor Claus, the *Phronima elongata*, lives in the same manner; but instead of occupying a living house, it generally seeks an empty lodging, in which it establishes itself like a pagurus.

The "Bernard the Hermit" of the Marseillaise fishermen, the *Pyades*, becomes the messmate of an anemone· which Dugès has called *Actinia parasitica*. According to the observations of the learned professor at Montpelier, the mouth of this anemone is always situated opposite to that of the crustacean, to take advantage of the morsels which escape from his pincers. Both of them profit by this association; and the opening of the shell is prolonged by a horny expansion furnished by the foot of the actinia.

On the coast of England lives another soldier-crab (*Pagurus Prideauxii*), which has as its principal messmate a sea anemone called *Adamsia*, which Mons. Greeff found at the island of Madeira. This pagurus is especially remarkable for the good understanding which exists between himself and his acolyte—he is a model Amphitryon. Lieut.-Col. Stuart Wortley has watched it in its private life, and thus relates the result of his observations : this animal after he has fished, never fails to offer the best morsels to his neighbour, and often during the day, ascertains if it is not hungry. But more especially when he is about to change his dwelling, does he redouble his care and his attention. He manœuvres with all the delicacy of which he is capable, to make the anemone change its shell ; he assists it in detaching itself, and if by chance the new dwelling is not to its taste, it seeks another until the *Adamsia* is perfectly satisfied. This association is not confined to the union of a decapod with a nereid and an actinia ; a curious cirrhipede often establishes itself on the body of the pagurus, and on the outside of the shell we generally find a colony of polyps, of a rose or yellow colour, which extend like a living carpet round this habitation. Thirty-six years ago we have given the name of *Hydractinia* to these polyps, which were till then entirely unknown to naturalists, and which form habitually a double overcoat for the paguri, if I may employ the expression of my learned colleague, Mons. Ch. Desmoulins.

In the Mediterranean lives the *Perella di mare* of the Italian fishermen, the *Reclus marin* of the Marseillaise ; this Alcyonium ought, by its manner of life, to be placed near the Hydractiniæ, and has been carefully studied by Mons. Ch. Desmoulins. It is the *Alcyonium (Suberites) domuncula* of Lamarck and Lamouroux.

The abdomen of these paguri is not only sheltered in a shell, but habitually visited by isopod crustaceans, described under the names of *Athelca, Prosthetes*, and *Phryxus,* which have entirely lost the livery of their order.

In the same association we also find the *Liriope*, a little isopod crustacean, of which much has been said, but which for a long time obstinately resisted all attempt at observation.

This latter personage is an isopod crustacean, of moderate size, which chooses the Peltogaster as a place of abode, after having undergone a very curious regressive metamorphosis. In fact, the young lyriope has at first its little feet like other isopods, but in the adult state, the female loses her antennæ, and changes her buccal as well as her branchial appendages, so as to assume a different appearance. Several naturalists have already endeavoured to give the life-history of this singular Bopyrian. The illustrious Rathke of Königsberg discovered it; Professor Lilljeborg, of the University of Upsal, gave the first account of it; and finally Professor Steenstrup of Copenhagen made known its true origin. In short, the Lyriopes are Bopyrian Isopods, living on cirrhipedes (Sacculinideæ) as real messmates, if not as parasites ; the male preserves his dignity and his prestige, but the female strips herself of all the attributes of her sex, and descends to the lowest degree of servitude.

Faujas de Saint-Fond has mentioned a fossil hermit-crab as found in the mountain, St. Pierre de Maestricht; but he called by this name a crustacean of the genus *Callianassa* and not a pagurus. These *Callianassæ* are always completely isolated in the chalk, and it is probable that they have no other domicile than the sand or ooze at the bottom of the sea, in which they hollow out galleries for themselves. Lobsters act in the same manner after moulting. The *Gebiæ* live like the Callianassæ, hidden in the mud. The *Limnaria lignorum* and the *Chelura terebrans* dig out a retreat for themselves in wood, like the Teredines.

We have just seen that the higher crustaceans, with their well-mounted eyes, their enormous antennæ, and their formidable pincers, are not all of them the great lords they pretend to be ; more than one of them has to hold out its hand and to accept humbly the assistance of its neighbours.

In the group of isopod crustaceans we find many necessitous beings, which, too proud to ask for food, are contented to take their place on some fish which is a good swimmer, which they abandon as soon as their interest demands it; if their host conducts them to regions that do not suit them, or if they have otherwise to complain of him, they give him up, and begin their maritime peregrinations with a fresh colleague. They always preserve all their fishing tackle and their sailing gear, and the female does not change her dress any more than the male. We have to notice that these crustaceans often identify themselves so entirely with their host that they seem to be a portion of him, and even to assume his peculiar colour. This is not a sign of servility, but a means of passing unobserved, and of escaping from the sight of the enemy that is watching them. Naturalists have given the name of *Anilocræ* to some of these free messmates.

Any one who has remained for some time on the coast of Brittany, especially at Concarneau, and who does not look with indifference on the many superb fishes which are taken every day, cannot fail to have been struck with the presence of a rather large crustacean, which clings to the sides of several kinds of *Labra*, especially the smaller species. This crustacean is an Anilocrian so common that we can scarcely imagine it to have escaped the attention of any naturalist. Nevertheless, no work makes mention of the regular attendance on the Labra by the Anilocra, which bears, we know not why, the specific name of *Mediterranean*. Rondelet was probably acquainted with it, when he spoke of the fish-lice, which do not derive their birth from these fishes, but from the sea mud. We often see males by the side of females on the same individual.

Some years ago a school of large cetaceans, known under the name of Grindewhalls or Globicephalæ were pursued in the Mediterranean, and those which were captured contained in the cavity of their nostrils, isopods closely allied to the *Cirolana spinipes*, if not identical with it. Till then the isopods had only been found on sea fishes; fresh-water fish are not, however, entirely exempt; in fact, a species of Œga (*Œga interrupta* of Martens) has just been found on the skin of a fresh-water fish of Borneo, the *Notopterus hypselonotus*. This same genus includes a species (*Œga spongiophila*) which lives in the magnificent sponge, the *Euplectella*. We know also a certain number of isopods which prefer the interior of their neighbour's body, and instal themselves in the cavity of the mouth, either to fish at the same time as their host, or to seize the food on its passage; others are of such a cruel nature, that they make no scruple to establish themselves in the stomach of a peaceable white fish. Without injuring any important organ, they penetrate in couples between the intestines, and, concealed in this retreat, they seize by the narrow entrance door, which they keep half open, all the little animals which are sufficiently bold to pass by. The cruelty of these beings knows no bounds. To instal themselves conveniently, they pierce the body of their host, skilfully open his stomach, and live there as Sybarites; their lodging is in future assured to them, and their fate is bound up with that of their host. Dr. Herklots, who has unfortunately been recently lost to science, communicated in 1869, to the Academy of the Netherlands, a very interesting memoir on two crustaceans of a new species, the *Epichtys giganteus*, which lives on a fish of the Indian Archipelago, and the *Ichthyoxenus Jellinghausii*, which lodges in a fresh-water fish of the Island of Java. It is to the latter that we refer here, and it seems that in this species we are approaching the limits at which commensalism commences.

The *Cymothoes* constitute another category of very interesting Isopods; they lodge with their female in the cavity of a fish's mouth. Dr. Bleeker, who has so successfully explored the Indian seas, obtained more than twenty species of these; but unfortunately he has not made a note of the fishes which harbour them. He has, however, made one exception with regard to a fish from the roadstead of Pondicherry, which is two feet long, and is called a Bat. It is known to naturalists under the name of *Stromatea Nigra;* its flesh is much esteemed, and it carried in its mouth a Cymothoe called by Dr. Bleeker *Cymothoe Stromatei*. A cymothoe has also been observed in the mouth of an Indian Chetodon. De Kay found one in a Rhombus in the United States, and De Saussure saw another at Cuba; and lately, Mons. Lafont discovered one in the Bay of Arcachon, on the *Boops*, and on the *Trachina vipera*. These cymothoes are about fifteen millimetres in length, and often fill all the cavity of the mouth. The most curious of all is that which is found in the mouth of the flying-fish, a kind of herring with elongated fins, which it uses as wings to rise into the air, when too closely pursued in the water. My son, when examining these fishes, in his passage from Cape Verd to Rio de Janeiro, found in the cavity of their mouth an enormous female, firmly wedged in the branchial arches, with its head inclined outwards, and the male, which was rather smaller, installed at her side. Their dwelling thus by pairs, as well as the entire conformation of the animal, plainly shows that these crustaceans make themselves at home, and live as true messmates. Cunningham has given them the name of *Ceratothoa exoceti*. A short time since, these Cymothoes were only known on marine fishes, but it appears from recent observations, that fresh-water fish are far from being exempt from them. Mons. Gertsfeld has recently noticed some on the *Cyprinus lacustris* of the river Amour, and another in the Rio Cadea in Brazil, on a *Chromida*. Other isopods also resort to fishes, and to animals of their own class, but they live as true parasites, and change their form as soon as they have chosen a resting-place. We shall return to this subject again. Some which are very common on prawns, are known under the name of Bopyrus.

An interesting division of amphipods have received the name of *Hyperinæ*. These crustaceans generally swim with facility, but walk with difficulty. They therefore usually have recourse to fishes, or even to medusæ, in order to gain support. We find on our own coasts the

Hyperina Latreillii, lodged in the superb *Rhizostoma,* which regularly appears in the later season of the year on the coast of Ostend; and a long time since, in 1776, O. F. Müller gave to a species of this genus the name of *Hyperina medusarum.* Mr. Alexander Agassiz once found a *Hyperina* on the disc of an *Aurelia.* The medusa, when extended, forms for them a balloon with its parachute, which supports and conveys them with greater or less rapidity. Professor Möbius has but lately remarked the presence of *Hyperina galba,* Mont., in the *Stomobrachium octocostatum,* Sars, a small species of medusa which appears in the Bay of Kiel in October and November. This naturalist supposes that these messmates at first inhabited the *Medusa aurita,* and then migrated into this species.

Besides these, there are *Gammari,* which, according to Semper, live in the *Avicula meleagrina* (pearl mussel), and are perhaps the principal manufacturers of fine pearls. The immense buccal cavity of the fishing-frog (*Lophius piscatorius*) is the abode in the Mediterranean of an *Apterychta,* and in the Northern Ocean of a curious amphipod of the ordinary size of the *Gammarus,* which takes a voyage without expense, and with no fear of wanting provisions. My son discovered it at Ostend, and proposes the name of *Lophiocola* to distinguish it. The Gammari give lodging themselves to a great quantity of parasites, which they must introduce into the bodies of those to whom they serve as food. It has been long known that whales have lice, to which naturalists have given the name of Cyami. They are found on the whales of both hemispheres, and on some other cetaceans. It is very remarkable that they are seen on the true whales of the north and of the temperate regions, on the *Megaptera*, and on several *Catodonta*, and that none are found in the *Balenoptera*. Mr. Dall has just noticed some on the singular *Grey Whale* of California. In general, we may say that each cetacean which harbours them, has its own species. Are they parasites or messmates? If we are to believe Roussel de Vauzème, they feed on the skin itself of the whale, the remains of which, it is said, are found in their stomach. According to this naturalist, the parts of the mouth are not adapted for suction, and the stomach contains ruminating apparatus. We think that a fresh examination is necessary before this question can be determined. The *Cyami* seem to us to live on the whale, as the *Arguli* and the *Caligi* do on fish; and if these living creatures derive their nourishment only from the mucous products secreted by the skin, we may ask whether they ought not to be classed in a separate category, for they ought not to figure on the list of paupers. We have found the orifice of the *Tubicinella* covered with cyami of every age, and their abundance in this place seems to indicate that their food was not supplied to them by the skin of their host. Mons. Ch. Lutken has recently published a very interesting monograph on these curious animals; according to him the *Cyamus rhytinæ*, which was thought to proceed from a piece of the skin of a *Stellerus*, appears to have been found on the skin of a whale.

The Picnogonons, the nature as well as the kind of life of which has been so long time problematical, deserve to be ranked among messmates, at least during their youth; in fact, after being hatched, they live on the *Corynes*, the *Hydractiniæ*, and other polyps, while at a later period they frequent molluscs or higher classes; Allman mentions the case of a *Phoxichilidium coccineum* lodged in a *Syncoryne*.

There are, perhaps, many other crustaceans which, placed among messmates, like the *Pandarus* and others, would have a right to claim a further inquiry. It is a fact that they are never seen except on the skin of their host, where they are always visible, preserve their colours entire, and never change their costume for the undress of a parasite. The *Pandari* live especially on the *Squalidæ*. Some which are found in our seas are of rare elegance of form. We must, perhaps, place among messmates the crustacean which Siebold found in the Adriatic, at Pola, on the belly of the worm *Sabella ventilabrum*, and it is not impossible that the *Staurosoma* observed by Will on an actinia, should have its place here rather than among the parasites.

A Rotifer without vibratory ciliæ, the *Balatro calvus* of Claparède, lives as an epizoon on the same annelids which lodge the Albertia in their interior. The Darwinists, observes Claparède, will not fail to remark the presence of these Rotifers of the genus Albertia in the interior of the animal, and of the genus Balatro on the exterior. The parasite Balatro, like a shadow, never quits his Mecænas, says the learned naturalist of Geneva; who has observed it on the *limicolous Oligochæts* of the Seime, in the Canton of Geneva.

The *Nebalia* of Geoffroy is an interesting crustacean, abundant on the coast of Brittany. This charming animal gives lodging habitually to a messmate which Mons. Hesse considered as an animal allied to the

Histriobdellæ, but which is only an imperfectly described Rotator. We believe that it is the same animal to which Professor Grube has given the name of *Seison nebalia*. It appears to assume the aspect of the Histriobdellæ, and may perhaps be adduced as an example of mimicry.

The molluscs, whatever their name may imply, are those which show the most independence among all the inferior ranks of animals ; not only are they contented with the slowness of their pace and the wretchedness of their food, but they only very rarely seek help from their neighbours. It is not, however, uncommon to find some living among corals, which have even been designated coralligenous molluscs. There exists a group of Gasteropods, the Eulimæ, which lodge in certain Echinoderms, and in every respect deserve to be classed among messmates; it was a long time before the relation which exists between them and the animals which shelter them had been thoroughly appreciated. Dr. Gräffe found one species, the *Eulima brevicula*, on the *Archaster typicus* of the Uvea Islands, in the Pacific Ocean. The molluscs, known by the name of *Stylifer*, have the same mode of life; they have been observed in the Asteriæ, the Ophiuræ, the Comatulæ, and even in the Holothuriæ ; and as they inhabit the digestive cavity of these animals, it was believed that they frequented them as parasites. This was the opinion expressed first by d'Orbigny, and adopted by most naturalists. Professor Semper found some in the skin of a holothurian (*Stichopus variegatus*), which he considered incapable of nourishing themselves otherwise than at the expense of their host. However this may be, these molluscs, ranged alternately among the *Phasianellæ*, the *Turritellæ*, the *Cerithia*, the *Pyramidellæ*, the *Scalariæ*, the *Rissoairia*, or in a distinct family, seem to belong rather to messmates than to parasites. We meet with Stylifers at the entrance of the mouth (Montacuta); more frequently they prefer, like the Fierasfers, to lodge themselves deeply in the digestive cavity in the midst of the *débris* of the prey. The Melania (*M. Cambessedesii*, Risso), which Delle Chiaie found in the Bay of Naples, on the foot of some comatulæ, belongs probably to this group of molluscs.

Among the gasteropod molluscs which are not able to maintain themselves, we may mention another, a curious parasite, which instals itself in one of the rays of a star-fish, and whose presence is revealed by a swelling which is not produced in the other rays. This mollusc has received the name of *Stylina*.

The molluscs which are the most remarkable from the point of view from which we are now considering them, are the *Entoconchæ;* they live in Enchinoderms, and it was thought for a while that we could see in them an example of the transformation of one class into another. Some years since J. Müller found in a Synapta from the Adriatic, tubes with male and female organs, without any other apparatus, and in these tubes appeared eggs, whence this great physiologist saw molluscs proceed, with a helicoid shell, similar to that of a small natica; he gave them the name of *Entoconcha mirabilis*. Professor Semper has since discovered another species of these, which he has dedicated to the illustrious physiologist of Berlin, and which he found attached to the cloacal sac of the *Holothuria edulis*.

The true relation between these molluscs and the holothurians remains to be discovered, and how the entoconchæ become at last simple sexual tubes. At present we must admit that it is the result of a retrogressive development like that of the peltogasters, which, like them, lose all the attributes of their class. They ought, perhaps, to be placed farther on, among parasites.

Some years since, some molluscs were observed which have compromised more or less the dignity of their class. Gräffe cites a species of the genus *Cypræa*, which one would certainly not expect to find in this category; it lives among the Viti Islands, in the compartments of the *Milithæa ochracea*. We have referred to it before. Naturalists have given the name of Melithæa to a very beautiful polyp which forms colonies of two or three metres in height. Mons. Steenstrup, with that perspicacity which discerns the most complex phenomena, has also described *Purpuræ* which live as messmates with the Antipathes and the Madrepores. Quite recently, indeed, Mr. Stimpson has observed in the port of Charleston, a gasteropod mollusc, similar to a Planorbis (*Cochliœlepsis parasitus*) which lives as a messmate in the body of an annelid (*Ocœtes lupina*).

It is not the same with a mollusc called *Magilus*, which naturalists considered for a long time to be the calcareous tube of an annelid. All conchologists know the shell of the Magili, so valued by collectors. This gasteropod when young takes up its lodgings in the substance of a madrepore which grows more quickly than he, and in order not to die, stifled in this living wall, he constructs a calcareous tube similar to the shell, of which it appears to be the continuation, and which allows it to procure for itself water, air, and food. The animal, protected by the madrepore, can do without its calcareous mantle, and only shows the end of the tube at the outside. It is this organ which sustains the struggle against the exuberant growth of the polyp, since it is by means of it that the mollusc obtains nourishment. The Magilus is like an oyster which is living in contact with a bank of mussels, with this difference, that the oyster almost always succumbs, while the magilus is always victorious in the struggle. We might also cite as well as the Magili, some *Vermeti*, certain *Crepidulæ* and *Hipponices*, which struggle with the same success against those which pilot or receive them.

As there exist parasites which only depend on others during their youth, so there are messmates which are completely independent when fully grown. Jacobson, of Copenhagen, wrote, in or about 1830, a memoir to show that the young bivalves which are found in the external branchial processes of the Anadontæ are parasites, and he proposed for them the name of *Glochidium*. Blainville and Duméril were charged to make a report on this memoir, which the author had sent to the Académie des Sciences. But his opinion had not many supporters, and it is now thoroughly known that the young anodonts differ considerably in their early and their full-grown state. During their stay in the branchial tubes, each young animal carries a long cable which descends from the middle of the foot, and serves to attach the anodont to the body of a fish, and yet permits it to move to a certain distance.* In fact the young anodonts have,

* I owe this observation to Dr. W. S. Kent, who showed me, in London, anodonts attached in this manner to sticklebacks.

not like the other acephala, vibratory wheels in order to move themselves; they are conveyed in this manner by their neighbours. There are also messmate acephala, as the *Modiolaria marmorata*, which lodge on the mantle of ascidians. Professor Semper found attached to the skin of a *Synapta similis*, a mollusc which possesses a peculiarity rare among these animals, that of carrying its shell in the interior and not on the outside.

There are few animals so infested with parasites as the Ascidians in general. Not only does their surface sometimes become a *microcosm*, as the name of one Mediterranean species indicates, but even in the substance of their testa lodge *Crenellæ* and other molluscs and polyps, which choose by preference to place their dwelling there. There are also Annelids which hollow out galleries in their interior, Lernæans which establish themselves in their respiratory cavity, Nematodes, Pycnogonidæ, Ophiuræ, and many others besides. Mons. Alfred Giard has described several Amphipods and Isopods which establish themselves on Tunicates. One cannot say that there is always such a complete agreement between animals of such different kinds, for Mons. Alfred Giard gives examples of grave disagreements which he has seen break out, and which have caused the death of several among them.

Another association is that of a gasteropod with one of the acephala. In the environs of Caracas lives an Ampullaria (*Crocostoma*) which lodges in the umbilicus of its shell another mollusc, the only fluviatile species of those countries, called the *Sphaerium modioliforme*. We have every reason to suppose that the Sphaerium lives on good terms with the Ampullaria, since they are usually found associated.

The Bryozoaria, the animal mosses, establish themselves on all solid bodies at the bottom of the sea, like true mosses on stones or on trees. One species, a *Membranipora*, is usually found on the common mussel. These animals are of small size, group themselves in colonies on the surface of shells and of polyparies, or even on crustaceans, and form by their union a fine kind of lace, the dazzling whiteness of which often comes out sharply on the varying and glittering colour of the shell. This is because each animal lodges in a cell which is not larger than the head of a pin, and all the cells of a colony are grouped together with the symmetrical regularity of the façade of a Gothic building.

Many Bryozoaria live in such a manner that it is impossible to say whether they are messmates, or have installed themselves by chance in a hiding-place for which they have no predilection. A charming bryozoon is developed in abundance on the carapace and the claws of the *Arcturus Baffini*, on the coast of Greenland, and propagates itself with extreme rapidity. On a single Arcturus we have found, scattered over its claws by the side of each other, Balani, Spirorbes, Sertulariæ, and vast colonies of Membranipora. One can see, merely by this example, the great zoological riches of the polar seas.

Certain annelids off the coasts of Normandy and Bretagne are the abodes of a bryozoary known under the name of *Pedicellina*, or *Loxosoma*. This interesting animal, which my fellow-labourer, Mons. Hesse, took for a Trematode, and whose drawings had led me into error, lives like others at liberty while young, and soon fixes itself to a Clymenian, in order to pass as a messmate the later period of its life. We have called it

Cyclatella annelidicola, because of its residence in a Clymenian annelid. Claparède and Keferstein have observed a
species, the *Loxosoma singulare,* on a capitellian annelid,
of the genus *Notomastus,* at St. Vaast-la-Hogue, on the
coast of Normandy. After this, Claparède found another
species, the *Loxosoma Kefersteinii,* in the bay of Naples,
on an *Acamarchis,* a bryozoarian mollusc. Mons.
Kowalewsky has observed in the Bay of Naples the
Loxosoma Napolitanum.

We found some years ago the Pedicellinæ in so
great abundance in the oyster beds of Ostend, that the
baskets and other things floating on the water were literally covered with them. We have several times since
endeavoured to procure them again, but it was in vain
to search in the same places where they were formerly so
abundant: we have not been able to discover a single one.

The class of worms includes not only parasites, it
contains also, as we shall see, true messmates; we find
some on crustaceans, on molluscs, on animals of their
own class, on Echinoderms, and on Polyps.

One of the most curious of these worms is the
Myzostoma, whose true nature has just been revealed by
the excellent researches of Mons. Mecznikow. These
myzostomes resemble trematode worms, but they have
symmetrical appendages, and are covered with vibratory
ciliæ. They live on the comatulæ, and run upon these
echinoderms with remarkable rapidity. They have not
hitherto been found elsewhere; they are evidently no
more parasites than the last mentioned, and their place
is among free messmates. Two great annelids are
found, the one, the *Nereis bilineata,* by the side of Paguri
in the same shell, the other, the *Nereis succinea,* accord-

ing to Grube, in the tubes or galleries of the Teredines. These dangerous acolytes introduce themselves furtively into the retreat of their host; and, always on the watch, they obtain at all times, and in every place, a certain prey, and a hiding-place from which they can take their share of their neighbour's goods. Another nereis, observed by Delle Chiaie, *Nereis tethycola*, lives in the cavities of a sponge, the *Tethya pyrifera*, which is visited by so many messmates and parasites, that it becomes a kind of hotel, where every one establishes himself at his ease. Rissò also mentions a *Lysidice erythrocephala* which lives in sponges.

In the same class is found an Amphinoma, a beautiful red-blooded worm, which proudly wears a plume of red branchiæ on its head, and which Fritz Müller observed on the coast of Brazil, begging assistance from a poor *Lepas anatifera*. Many Polynoës live upon other annelids; the *Harmothoë Malmgreni* on the sheath of the *Chætopterus insignis*, the *Antinoe nobilis* on the case of the *Terebella nebulosa*. Prof. Ray Lankester has lately communicated some observations on this subject to the Linnæan Society of London, and Dr. M'Intosh mentions some new species leading the same kind of life on the coast of Scotland.

Grube found at Trieste, in a star-fish (*Astropecten aurantiacus*), between its rows of suckers, a *Polynoë malleata*, with its stomach attached to the animal; and Delle Chiaie has lately observed on an asteria, a *Nereis squamosa* by the side of a *Nereis flexuosa*. Mons. Grube thinks that the nereis of Delle Chiaie is no other than the *Polynoë malleata*. Lobsters are often covered with very small tubicular worms, which invade the whole

4

carapace, and which, as true messmates, give themselves up to the caprices of their host. These are a kind of *Spirorbis*, which, under the form of small spiral tubes, instal themselves, by preference, on the limbs, the antennæ, or the claws.

Mr. A. Agassiz has seen on the coast of the United States, a Beroë (*Mnemiopsis Leidyi*) which gives lodging in its interior to worms which somewhat resemble the Hirudinidæ, and which doubtless live there as messmates. Mr. A. Agassiz has remarked to me another example of commensalism. On the coast of the territory of Washington, as far as California, is found a worm of the genus *Lepidonotus*, which always lives near the mouth of a star-fish, the *Asteracanthion ochraceus* of Brandt; sometimes as many as five are found together on a single individual, and are placed on different parts of the ambulacral rays. Mr. Pourtalis and Mr. Verril have observed annelids lodged in the polypidoms of the Stylaster.

There are few fish on which are not found *Caligi*, charming crustaceans which please the eye by their attenuated shape and their graceful movements. On these Caligi, which sometimes literally cover the skin of cod-fish coming from the north, we often find a curious trematode, the *Udonella*, which resembles one of the small hirudinidæ. Should this worm be placed among messmates? What is the part which it plays? We are persuaded that it is the same as that of the histriobdellæ under the tail of lobsters, that is to say, that it clears off the eggs of caligi which do not arrive at perfection, but perish in the course of their evolution.

Roussel de Vauzème has mentioned another worm, a

nematode, to which he has given the name of *Odontobius*, and which lives on the palatal membranes (the whalebones) of the southern whale. It is evidently a messmate. It can get nothing from the whalebones, but it snaps up on their passage in the interstices of the baleen, small animals of all kinds which swarm in these waters. When we open the *Pylidium girans*, we often find in the interior of its digestive cavity a larva, which was once thought to be descended from it, but instead of being allied to the Pylidium, this larva comes from a nemertian known by the name of *Alardus caudatus*. The young nemertian never abandons his host until it approaches the period of puberty, and then all the individuals living under the same conditions emancipate themselves at once, to pass the rest of their days free and roving like their mother.

Worms which have less freedom, like the Distomians, are sometimes both messmates and parasites. We find a remarkable example of this in the *Distomum ocreatum* of the Baltic. According to the observations of Willemoes-Suhm, this trematode passes its cercarial life freely in the sea, and instead of encysting itself in the body of a neighbour, it attaches itself to a copepod crustacean, the whole of the inside of which it devours, in order to clothe itself afterwards with the carapace of its victim. It is under the cover of its prey that it passes into the herring, and completes its sexual evolution..

Mons. Ulianin has recently found another Distome (*Distomum ventricosum*) which passes its cercarial life in freedom in the bay of Sebastopol, and completes its evolution in the fishes of the Black Sea. J. Müller has long since found Cercaria living freely in the Mediterranean.

We ourselves, some years ago, while making some researches among the Turbellaria, found among the eggs of some ordinary crabs of our coasts (*Carcinus mænas*), an interesting worm which we named *Polia involuta*, but which Prof. Kolliker appears to have known before, and designated by the name of *Nemertes carcinophilus*. It is not known whether it plays the same part as the Histriobdellæ and the Udonellæ. Delle Chiaie, as well as Prof. Frey and Prof. Leuckart, make mention of another nemertian which inhabits the *Ascidia mamillata*. Among the nemertians, we may allude to the *Anoplodium parasita*, which lives in the *Holothuria tubulosa*, and the *Anoplodium Schneiderii*, inhabiting the intestines of the *Stichopus variegatus*.

According to Mr. A. Agassiz, a species of Planarian (*Planaria angulata*, Mull.), lives as a free messmate on the lower surface of the Limulus, and prefers to establish itself near the base of the tail. Mons. Max Schultze recognized last year this same messmate on a limulus, which had died at Cologne in the large aquarium, and which had been sent to him for his anatomical studies. He showed at the congress of German naturalists at Wiesbaden, in 1873, the drawing which he had made of this animal, which he thought new to science. We may remark in passing, that he arrived, by means of his anatomical observations on Limuli, at the same result as did my son by his embryogenic observations, namely, that these supposed crustaceans are to be regarded as aquatic scorpions. Mr. Leidy also makes mention of Planarian parasites (*Bdellura*), with a sucker at the extremity of the body; and Mons. Giard noticed a blue one on the body of a Botryllus.

But of all the Turbellaria, the genus which appears to us the most interesting is the Temnophila, which Gay first observed on crabs at Chili, and which Professor Semper afterwards found on the crabs of the Philippine Islands. Gay and Phillipi found colonies of these animals on the body, the claws, and more especially the abdomen, of the *Œglea*. This messmate resembles a trematode by its form and by its posterior sucker, but by its entire character, and especially by its sexual organs, it belongs to the *Turbellariæ*. Mons. Blanchard calls it *Temnophila Chilensis*. Professor Semper saw at the Philippine Islands these Temnophilæ on river crabs, at five thousand feet above the level of the sea.

The *Cydippe densa*, a charming polyp of the Gulf of Naples, lodges in its gastro-vascular apparatus larvæ of annelids, which may as well be considered parasites as messmates. We owe to Panceri the first observations on these worms, of which two genera, *Alciopina* and *Rhynconerulla*, seem to live in the same manner in their youth. A naturalist, whose loss is profoundly deplored by the scientific world, Claparède, occupied himself with observations on these annelids during the last years of his life. It appears that these worms are so common in these polyps, that four have been found at once in the same animal.

The Spoon-worm, named by Œrsted, *Sipunculus concharum*, ought doubtless to find its place here. An oligochete worm, *Hemidasys agaso*, from the Gulf of Naples, lives on the *Nereilepas caudata*, and Claparède did not think it unworthy of his attention. The surest means of finding it, says this philosopher, is to look for it on this annelid; and our much regretted fellow-labourer at Geneva did not abandon this messmate before he had completely studied it. Let us remark in passing, that Professor Grube published in 1831, at Königsberg, a special work on the abodes of annelids in general.

Cases of commensalism among the Echinodermata are still more rare. These animals are sufficiently provided with organs, both with respect to their food and their skin, not to require the assistance of their neighbours. We cannot rank as a phenomenon of commensalism, the conduct of the young Comatulæ, which fasten themselves, as Mr. A. Agassiz informs me, to the basal cirrhi of the adult echinoderms, and there form a little colony of young Pentacrinites.

We only know one Ophiurus (*Ophiocnemis obscura*), which lives as a messmate on a comatula, and consequently seeks assistance from an animal of its own rank. Another kind of Ophiuride (*Asteromorpha lævis*, Lym.) fixes itself on a *Gorgonella Guadelupensis* of Barbadoes. Everything induces us to suppose that we shall find more than one species of echinoderm, which will take its place among these when their mode of life has been studied with greater care. Professor Lütken has just proved this by quite recently making known another *Ophiothela*, which lives in the straits of Formosa, and seems to be the messmate of an Isidian polyp, known under the name of *Parisis loxa*. Another species (*Oph. mirabilis*) from Panama, infests certain Gorgoniæ and sponges ; a third is found in the Fiji Islands on the *Melitodes virgata ;* a fourth at the Isle of France on Gorgoniæ ; and a fifth at Japan on the *Mopsella Japonica.* There is also another in the Pacific Ocean, but its companion is not known.

Professor Mobius, as well as Dr. F. Martens, has noticed a *Hemieuryale pustulata* on a polyp of Jamaica, known under the name of *Verrucella Guadelupensis*. This is a curious instance of mimicry.

The class of polyps includes several species which seek for assistance from others, and are classed among messmates. One of the most remarkable is the Gigantic Medusa, which can extend its arms downwards to a hundred and twenty feet, and bears the name of *Cyanea arctica;* the disc is seven feet and a half in diameter, and when the animal is on the surface of the water, the fringes, which surround the cavity at its mouth, occasionally afford lodging in the midst of them to a species of actinia, which lives there as messmate. Sometimes three, and even four or five, are found on a single Cyanæa. This also is an observation due to Mr. A. Agassiz, which he has published in his interesting work, "Sea-side Studies." Prof. Haeckel supposed that the *Geryoniæ* produce *Œginidæ* by means of buds; but it appears that the learned professor was mistaken as to the nature of these buds; that instead of being produced one from the other, they have, according to Steenstrup, a completely different genealogy, being only united by conditions of good-fellowship. They may be truly called messmates.

Mons. Lacaze-Duthiers, who went to the coast of Africa to study corals, met with a young polyp which requires the assistance of another polyp in its early condition. This animal, to which he has given the name of *Gerardia Lamarckii*, lives on one of the Gorgoniæ, which it invades and stifles, as the lianas strangle the tree over which they spread themselves. But these same Gerardiæ can also develop themselves on the eggs of the *Plagiostoma*, and are then capable of living separately. In the substance of this polyp lives a crustacean, the nature of which Mons. Lacaze-Duthiers has not yet made known.

The superb sponge, *Euplectella aspergillum*, the elegant structure of which cannot be sufficiently admired, is, unlike the Alcyonium of the Dromia, rooted to the soil, but nevertheless gives shelter to three kinds of crustaceans : Pinnotheres, Palemonidæ, and Isopods. These supposed plants have been known for many years under the Spanish name of *Regadera*, or the English "Venus' Flower-basket;" they were first brought from Japan, and afterwards from the Moluccas, and more recently from the Philippine Islands In almost all the individuals which Professor Semper was able to study in those parts, were found the same crustaceans. These *Euplectellæ* have just been met with to the south-west of Cape St. Vincent, by Wyville Thomson, who has brought up some from a depth of 1090 fathoms, while on board the *Challenger*.   This skilful professor has discovered another sponge to the north-west of Scotland, at a depth of 460 fathoms; it bears the name of *Holtenia Carpenteri;* and I have in my possession a fine specimen which I owe to his generosity, and keep as a *souvenir* of the delightful hospitality which he extended to me at the Edinburgh meeting.

There are also sponges which construct a dwelling in the abode of their neighbour. We find, among others, a small sponge known under the name of *Clione*, which establishes itself in the substance of the shell of oysters, and hollows out galleries as the teredo does in wood.

Mr. Albany Hancock found twelve species of Clione on a single Tridacna. They are evidently not parasites, and I am not sure if their place is properly among messmates. The oyster, and more especially the *Ostrea hippopus*, lodges three or four different sorts in its shell. These Cliones possess siliceous spicules, by means of which they hollow out galleries in the substance of shells. Mr. Hancock has published a monograph of this genus, in which he recognizes twenty-four species collected from different shells, and two other species, which he refers to the genus *Thoasa*.

The cliones are real lodgers which lead us to the *Saxicavæ*, the *Pholades*, and the *Teredines;* they seek their lodging in rocks or in wood; these lead directly to the sea-urchins, which also hollow out lodgings in rocks, but without penetrating deeply. Professor Allman has just observed a very remarkable case of commensalism between a sponge and one of the tubulariæ. The crown of the tubularia is extended at the entrance of the canals of the sponge; and the association is so complete, that the Edinburgh professor imagined that he had before his eyes a true sponge with the arms of a tubularia.

In the lowest ranks of the animal scale, there are certain kinds of animalcules, which establish themselves on the bodies of obliging neighbours, and take advantage of their fins in order to swim at their expense. Thus we often find the bodies of certain crustaceans covered with a forest of vorticellæ and other infusoria. They cause themselves to be towed like cirrhipedes, but they do not change their toilet like them, so that it cannot be said that they put on the livery of servitude.

The kind of life led by several of these animalculæ is as yet little known.

Mons. Leydig has found in the stomach of the *Hydatina Senta* a messmate which much resembles an Euglena, and still more the *Distigma tenax*, Ehr.

CHAPTER III.

FIXED MESSMATES.

THE animals of which we have just spoken usually preserve their full and entire independence; from the time of their leaving the egg, till their complete development, they are subject to no other outward changes than such as belong to their class. If they sometimes renounce their liberty, it is only for a limited time; and they all preserve not only their peculiar appearance, but their organs intended for fishing or for locomotion. It is not thus with those which we are now about to consider; they are free in their youth, but as they draw near to puberty they make choice of a host, instal themselves within him, and completely lose their former appearance : not only do they throw aside their oars and their pincers, but they cease sometimes to keep up any communication with the outer world, and even give up the most precious organs of animal life, not even excepting those of the senses; they are installed for life, and their fate is bound up with the host which gives them shelter. The number of these messmates is considerable.

We shall first allude to some crustaceans named Cirrhipedes by Lamarck. The metamorphoses which they have undergone since they left the egg have so

much changed them, that Cuvier and all the zoologists
of his age placed them in the class of mollusca. The
incrustations of their skin resembled shells, which these
creatures generally carry in the substance of their
mantle.

These ambiguous creatures are far from being microscopic; there are Balani which attain the size of a
walnut, and some have been found not less than ten
inches high, as the *Balanus psittacus.* Some years since
we saw on a piece of floating wood, found by fishermen
in the North Sea, Anatifæ on the end of stalks from
six to seven feet in length. The anatifæ themselves
were of the usual size. These cirrhipedes belonged to
every geological period; they have already been found
in the Silurian formation, but, unlike the trilobites their
contemporaries, they pass through all the ages, and, far
from decreasing, they reign as masters at the present
time in the two hemispheres.

It was an English naturalist, Thomson, who first
made known the true nature of these singular organisms. So far were many from understanding their
affinities with the other classes, that even after the
excellent researches of the Belfast naturalist, they
doubted their correctness, and supposed that these
animals were allied both to the mollusca and to the
articulata.

We see by this the immense progress which embryo-
logical studies have caused us to make in the appreciation of natural affinities. No one at the present time,
who has seen a cirrhipede hatched, can retain any doubt
as to the place which it ought to occupy. These crustaceans, taken as a whole, lead a life in which we find

more than one contrast; all live as wanderers when they first leave the egg, and they are hatched in such abundance on the coast, that the water becomes literally troubled with them. At the first period of their life, they have a supple and elegant body, and fins admirably divided, and the gracefulness of the postures which they assume does not yield in beauty to those of the most brilliant insect. After having spent some time in seeking adventures, they are seized with disgust for a nomad life; they choose a resting-place, and establish themselves by means of a cable which they afterwards abandon, and shelter themselves in an enclosed retreat for the rest of their days. Many cirrhipedes choose the back of a whale or the fin of a shark, and make the passage across the Atlantic or the Pacific in less time than the swiftest steamboats.

In many of these, recurrent development (I was about to say degradation) sometimes proceeds so far, that their animal nature becomes doubtful, and more than one of them, having no longer any mouth by which to feed, are reduced to a mere case which shelters their progeny. The messmate very nearly takes its rank among parasites. There are also cirrhipedes which live on different genera of their own family; and some species which are always found in society with other species. Some also live as messmates with each other; some of the Sabelliphili have one of the sexes parasitical on the other sex.

Crustaceans are usually dioecious; but because of their manner of life, the cirrhipedes sometimes unite the two sexes and thus render the preservation of the species more certain. The whole family of the *Abdominalia*, a name proposed by Darwin, if I am not mistaken, have the sexes separate; and the males, comparatively very small, are attached to the body of each female. It is a case of polyandria which we see realized in the *Scalpellum*. Darwin made known the existence of supplementary males, so small and so little developed, that they are with difficulty discovered, and so badly are they provided with organs that they have neither those of motion nor a stomach to digest. We have not exhausted the strange peculiarities of this particular group; there are some which live without shells and claws in the inside of other cirrhipedes, and atrophied males which only exist at the expense of their own females.

It is almost useless to make the remark that more especially here there exist almost insensible gradations of difference between parasites, messmates, and free animals, and we shall find more than one example of this in the crustaceans to which we now allude.

The most interesting fixed messmates are evidently those cirrhipedes, which, under the name of *Tubicinella*, *Diadema*, or *Coronula*, cover the skins of whales. They are, like all the rest, free in their infancy, but soon they take shelter on the back or on the head of one of these huge cetaceans, which they never quit when they have once chosen their abode. That which gives them great importance is, that each whale lodges a particular species; so that the crustacean messmate is a true flag which indicates in some respect the nationality, and it would not be without interest for voyagers who are naturalists to study these living flags.

The great whale of the north, the *Mysticetus*, which our northern neighbours discovered while seeking for an eastern passage to India, a species which never leaves the ice, carries no cirrhipedes. This fact was already known to Iceland fishermen of the twelfth century. The intrepid whalers of these regions used to distinguish a northern whale, without "calcareous plates," from a southern whale with plates, that is to say, with cirrhipedes. This latter whale is the celebrated species of temperate regions, the *Nord-Kaper* which the Basques used to hunt, from the sixth century, in the Channel, and which they used afterwards to pursue even to Newfoundland. The whales of the southern hemisphere, like those of the Pacific Ocean, all have their own species of cirrhipedes. We found in the museum of the Zoological Garden at Amsterdam, a *Coronula*, brought from Japan by Mr. Blomhof, known under the name of *Coronulæ reginæ*, which, no doubt, characterizes the whale of those latitudes. Another northern whale, the *Keporkak* of the Greenlanders, very remarkable for its long fins, which give it the name of *Megaptera*, is covered very early in its life with these crustaceans, so much so, that the Greenlanders imagine that they are born with them. Some even have pretended to have seen Megapteræ covered with these coronulæ before their birth. Eschricht has in vain offered a reward to him who would send him coronulæ still attached to the umbilical cord; he has only received some pieces of skin covered with hairy bulbs. There is no doubt that young whales have been seen and captured while following their mother, which were already covered by these crustaceans.

Steenstrup has indicated the presence of *Platycyamus Thompsoni* on the body of the *Hyperoodons*, and the *Xenobalanus globicipitis* on the globiceps of the Shetland Isles. The *Cryptolepas* is a new genus of Coronulidæ which inhabits the coast of California, on the singular mysticete recently distinguished by the name of *Rhachianectes glaucus*. The *Platylepas bisexlobata* has lately been observed on one of the Sirenia, the *Manatus latirostris*. The marine turtles are also invaded by these singular animals, and their peculiar form, joined to their habitat, has given them the name of *Chelonobia*. It is not uncommon to find by the side of these Chelonobiæ, and even upon them, the Tanaïs, Serpulæ, and Bryozoariæ, forming together an animal forest on the cuirass of the turtle. The *Matamata*, a turtle living in the brackish water of Guiana, is covered with a cirrhipede more allied to the ordinary balani than to the chelonobiæ. Other living reptiles are not more exempt from cirrhipedes than turtles; the *Dichelaspis pellucida* and the *Conchoderma Hunteri* invade different sea-snakes. Many sharks harbour particular kinds, among which we mention the *Alepas* of the *Spinax niger* from the coasts of Norway. The same Alepas has been found on the *Squalus glacialis* at the same time as the *Anelasma squalicola*. Half a dozen varieties of these are known, one of which inhabits an echinoderm, another a decapod crustacean. These kinds of alepas are so reduced when they are adult, and are so completely despoiled of their distinctive attributes, that it is necessary to study them with especial care in their first dress, in order to recognize their parentage.

Other cirrhipedes establish themselves on neighbours of their own class, and we also find crustaceans upon other crustaceans. A pretty genus lives near Cape Verd on the carapace of a large lobster, and spreads itself on the centre of the back like a bouquet of flowers. My son has procured some very fine specimens, an account of which he will publish, together with the other materials which he has collected during his passage across the Atlantic. Mr. John Denis Macdonald found in abundance on the branchiæ of a crab in Australia, the *Neptunus pelagicus*, which he places between the Lepas and the Dichelaspis.

The most singular, if not the most interesting of all these cirrhipedes, are the Gallæ, which appear under the tail of crabs or the abdomen of paguri, and which zoologists designate under the names of *Peltogaster* or *Sacculina*. They are found in both hemispheres. The recurrent development is so complete, that we can no longer distinguish any organic apparatus unless it be that of reproduction, and the whole body is a mere case enclosing within its walls eggs and spermatozoids. We see them very frequently under the abdomen of the crabs of our coasts, or even on the segments of the bodies of paguri. Mons. A. Giard has lately studied these animals. It is during the coupling season, according to him, that the Peltogasters establish themselves upon the crabs. Professor Semper has brought back quite a collection of them from his voyage to the Philippine Islands, and has entrusted them to one of his pupils, Dr. Kussmann, for the purposes of study. We heard him with great interest, at the late Congress at Wiesbaden, explain with remarkable clearness the results of his learned and conscientious observations. We do not think that we shall be wrong in adding that, for a long time, we shall see nothing better or more complete on this subject. All those cirrhipedes which adhere by their head to the skin of their host, by means of filaments, are now designated by the name of *Rhizocephala*.

A curious opinion, quite recently expressed by a naturalist, Mons. Giard, and which is a sign of the times, is that the Peltogaster of the Pagurus has become a Sacculina on the crab; the host having been transformed, its acolyte has done the same thing under the same influence.

Professor Semper has also found among the Philippine Islands, isopod crustaceans living as messmates after the manner of the peltogasters. Two cirrhipedes of the family of Peltogaster, the *Sylon Hippolytes* and the *Sylon Pandali*, have been found by Mons. Sars under the abdomen of the *Pandalus brevirostris*.

There are cirrhipedes on the gasteropod molluscs. The *Concholepas Peruviana*, that beautiful shell which has long been considered a rarity in our collections, is frequented by the *Cryptophiolus minutus*, only a sixth of an inch in length. The *Scalpella* often inhabit the Sertulariæ and other polyps; *Oxynasps, Creusiæ, Pyrgomæ*, and *Lithotryæ* inhabit corals. Certain kinds of sponges are regularly invaded by the *Acastæ* of Leach, eight species of which are mentioned by Darwin. As we find elsewhere parasites on parasites, here also we find messmates on messmates; on the common anatifa we perceive other genera, and on the *Diadema* of the North Pacific, we almost always see *Otions* and *Cineras*. The *Protolepas bivincta* also, a fifth of an inch in length, lives as a messmate in the mouth of the *Alepas cornuta;* and the *Elminius* of Leach also inhabits other cirrhipedes. The *Hemioniscus balani*, which Goodsir had taken some years ago for the male of the Balanus, is a messmate on these cirrhipedes. Parasites also are found in messmates; the soldier-crab gives lodging to the sexual

Eustoma truncata in its interior. A macrourous crustacean which we ought to mention here, the *Galathea spinirostris,* Dana, frequents a comatula, the colour of which it assumes; it is the same without doubt with the *Pisa Styx,* which lives on a polyp known by the name of *Melitæa ochracea.*

If we pass from the crustaceans to the molluscs, we have to notice in the first place an elegant gasteropod, the *Phyllirhoa bucephala,* which carries on its head a singular appendage, the nature of which has only lately been known; J. Müller took it at first for a medusa, then he abandoned this opinion, when at length Mons. Krohn referred it definitively to the lower polyps; it differs from its congeners only by its form, its tentacular cirrhi, and its mode of life: it is the *Mnestra parasites.* There are a great number of acephalous molluscs, which we might mention as messmates, but we will only refer to the *Crenellæ* which are regularly found in the substance of sponges.

The *Philomedusa Vogtii* of Fr. Müller, which·lives on the *Haleampa Fultoni,* undoubtedly deserves to be mentioned here as a fixed messmate. Many bryozoa spread themselves over marine animals, and often engage in a deadly struggle with their patron. But among all these bryozoa we must mention an animal very common on the sea-shore at Ostend, and which one would take for a dried leaf, the *Flustra membranacca.* On the surface of these imitative leaves are found little bouquets of other bryozoa, which are either *Crisiæ* or *Scrupocellariæ.* Another kind, which has also passed for a gelatinous plant, bears the name of *Halodactylus.* Without any micro - scopic study, one can obtain an idea of these colonies.

One of these Halodactyles spreads itself upon the stalk of a Sertularia, all the inhabitants of which it stifles, so that it is the victim himself who serves as a guardian to the invader.

These Halodactyli are very widely spread over the Northern Seas, and often establish themselves on the large horse-hoof oyster.   Michelin has noticed under the name of parasite a fossil *cellepore* from the saltpits of Touraine and Anjou, which entirely surrounds the shell of a gasteropod ; in order to prevent its patron from dying of hunger, the bryozoon develops itself around the mouth like a gallery, and prolongs its last spiral.   This *Cellepora parasitica* has evidently a place here.

Many of these messmate bryozoa are found in a fossil state in the crag of the Antwerp basin.

We have still to mention among fixed messmates many polyps, some of which are very remarkable.   Thus, many naturalists speak of vast colonies of polyps in which lodge various animals which shelter themselves there like paguri in deserted shells.

Among these are the colonies of which Forster speaks, which are not less than three feet in diameter, and fifteen feet in height, with a crown of eighteen feet in diameter.   Dana also makes mention of an *Astræa* of twelve feet in height, and of *Porites* twenty feet high, which contain more than five millions of individuals, among which a number of animals come to take refuge.

The Museum of Natural History at Paris is in possession of a superb specimen of *Porites conglomerata :* in the middle of the colony lodges a Tridacna (*Trid. corallicola*, Val.) like a pagurus under a forest of hydractiniæ.   This remarkable polyp was brought from the Seychelles Islands by Mons. L. Rousseau. It is not impossible that pinnotheres live in this same tridacna, and that we have there a fresh example of messmate within messmate.

In the Bay of Massachusetts, on the coast of New England, another curious messmate lives at great depths; Dana has lately described it, under the name of *Epizoanthus Americanus*, V. It establishes itself in the *Eupagurus pubescens*. The *Sertularia parasitica* of the gulf of Naples, from which I have formed the genus *Corydendrium*, is a messmate after the manner of an infinite number of other polyps. In closing this list, we shall mention a polyp, named *Halicondria suberea*, and the *Actinia carcinopodus* of Otto, which inhabit an univalve mollusc; as also the *Heterosammiæ* and the *Heterocyathi* of the family of Turbinolidæ, which lodge in a trochoid shell.

The sponges, placed by naturalists by turns among plants or on the confines of the animal kingdom, are now generally regarded as polyps; this is the opinion expressed by Haeckel, who wishes at the same time to replace the term Cœlenterata by that of Zoophytes. The learned naturalist of Jena, when making this proposition, should have remembered that in 1859 we placed the sponges in the group of polyps, as the lowest in the scale; and that we proposed, from the time when the acalephæ were recognized to be adult polyps, to designate all these animals under the name of Polyps. Some time after, R. Leuckart proposed the appellation Cœlenterate Polyps, which has been generally received. Professor Haeckel would have lost nothing by acknowledging that in 1873 he arrived at a result similar to

that to which I had come twenty years before, and that
it is not a very happy innovation to change the term
polyps for zoophytes. It is the more surprising that this
naturalist has forgotten to quote my opinion, since at the
congress of naturalists at Hanover in 1866, I had placed
this question on the agenda for an ordinary meeting.

I maintained, in opposition to the opinion of the
naturalists whose authority had been especially recognized in the matter (Osc. Schmidt, who was present,
among others), that sponges are lower polyps, whether
they are regarded as to their development or their
organization.

This group, so remarkable in form, so varied in
colour and appearance, very often affords examples of
animals which live with them as true messmates; and
we find the same relations established between them in
both hemispheres. As we observe rhizophales on crabs
and soldier-crabs, and pinnotheres on bivalve molluscs,
so we find that the sponges of the Indian Seas or of
Japan harbour the same messmates which we discover
on them in the Northern Seas or the Atlantic.

In the sea of Japan is found a very remarkable
sponge, generally known by the name of *Hyalonema*.
It is a bundle of spicules like threads of glass, which
seem artificially tied together, and on the surface of
which we regularly find a polyp of the genus *Polythoa*.
The nature of this sponge, and its relations with the
polyps which surround it, have been discussed for many
years. Ehrenberg had recognized the polyp Polythoa
around the spicules, but the Hyalonema was considered
by him as an artificial product. The *Polythoæ* were
regarded as only a case in which had been placed this

bundle of spicules. The learned microscopist of Berlin had even thought that he had found the proof of this opinion in the presence of woollen threads which were observed in a specimen which Mons. Barbosa du Bocage had sent him from Lisbon. Woollen threads had indeed adhered to the spicules of Hyalonema, but they came from the fishermen, who, when they drew these sponges from the water, placed them carefully in their bosoms under their woollen jerseys.

Dr. Gray, of the British Museum, considers the sponge as a parasite of the Polythoa, and that the bundle of spicules belongs, not to the sponge, but to the polyp. The most learned naturalist on the subject of sponges, Mr. Bowerbank, expresses a different opinion. The sponge and its spicules, according to him, are but a single body, and the polyps are only a part of it. The supposed polyps would only form a cloacal system for the use of the sponge colony.

Valenciennes, guided no doubt by the observations of Philippe Poteau, was the first to recognise the nature of the sponge and its spicules, but it is to Max Schultze that we must give the credit of distinguishing the true character of this extraordinary marine production. He has shown that the bundle is formed by the extraordinarily long spicules of the sponge, and that the polyp establishes itself upon it, by forming a sheath around the bundle.

The fact is no longer doubted by any one, that the long spicules form part of the sponge, and that the polyp establishes itself on a part of the colony. But science rarely advances by a single stride, and Max Schultze, like his predecessors, mistook the top of the sponge for the bottom; Professor Loven has shown the true pose of the Hyalonema, and this he has effected by means of a small specimen from the Northern Sea.

Semper found a new Œga, to which he gave the specific name of *Hirsuta*, in an enlarged canal of the new Hyalonema of the Philippine Islands, which he dedicated to Mons. Schultze.

The Adriatic also produces a species of the same genus (*Polythoa*) which inhabits, like that of the Chinese Sea, a sponge to which the name of *Axinella* has been given. These Polythoæ are only found on the Axinellæ, says Osc. Schmidt, who has especially studied the sponges of this sea and of the Mediterranean. Professor Gill mentioned at the last meeting of the scientific congress at Portland (1873), a new Hyalonema found on the coast of North America by the fishery commission of the United States. A memoir on these sponges, interesting in a systematic point of view, is due to the pens of Herklots and of Marshall.

We think that we ought to place among fixed mess-mates a very problematical organism which lives on Ser-tulariæ, especially on the *Sertularia abietina*, and which Strethill Wright has designated by the name of *Core-thria sertularia*. Claparède has given to this singular animal the more expressive name of *Ophiodendrum abie-tinum*.

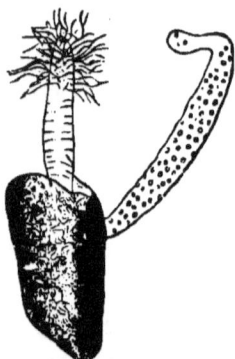

Fig. 1.–Ophiodendrum abietinum on Sertularia abietina.

We have regularly found it on the *Sertularia abietina* at Ostend, every time that we have had an opportunity of observing these polyps immediately that they have been raised from the bottom of the sea. It is an organism whose affinities are not yet established.

5

CHAPTER IV.

MUTUALISTS.

IN this chapter we bring together animals which live on each other, without being either parasites or messmates; many of them are towed along by others; some render each other mutual services, others again take advantage of some assistance which their companions can give them; some afford each other an asylum, and some are found which have sympathetic bonds which always draw them together. They are usually confounded with parasites or messmates.

Many insects shelter themselves in the fur of the mammalia, or in the down of birds, and remove from the hair and the feathers the pellicle and epidermal *débris* which encumber them. At the same time they minister to the outward appearance of their host, and are of great utility to him in a hygienic point of view.

Those which live in the water have other guardians: instead of insects, we find a number of crustaceans which establish themselves on fishes, and if there are no scales of the epidermis which annoy them, there are mucosities which are incessantly renewed in order to protect the skin from the continual action of the water.

We find many on the surface of the scales, and others which conceal themselves at the bottom of mucous canals. We have brought together only a few examples, and there are certain others which are mentioned elsewhere, but which ought more properly to be placed here.

The insects long known under the name of *Ricini*, and to which many other appellations have been given, deserve to figure in the first rank in this group. They have always perplexed entomologists, who seem to consider them as parasites allied to acaridæ and lice. It has, however, been long known that they have no trunk to suck with, and that they have two small scaly teeth, which rather serve for the purpose of biting. A long time since, the examination of their stomach proved that they contain only morsels of skin instead of blood. This has induced many entomologists to place them in the same order as grasshoppers, that of Orthoptera.

Lyonet has given figures of several of those which he studied with the care which he so well knew how to employ in his anatomical investigations; and in 1818 Nitzsch, a professor at Göttingen, had brought together so great a number of them, that it required several days to examine his collection; he began the publication of his catalogue, but has not had time to finish it. Several other entomologists and anatomists have since taken up the subject.

We owe the description of several hundred species to Mr. Denny. Mons. F. Rudow has lately made known a great number of species which he has collected from the skins of birds coming from Japan, Australia, Africa, and the two Americas.

Professor Grube, of Breslau, has published the description of the insects and acaridæ found during the travels of Middendorf in Siberia. These descriptions relate especially to the Philopteræ of birds, the Pediculinæ of the mammalia, a flea of the *Mustela Siberica*, and an acarus of the *Lemmus*. Quite recently, an American naturalist, Mr. Packard, who has undertaken the study of so many different subjects, has published in the "American Naturalist" the description, accompanied by an engraving, of the *Menopon picicola*, found on the *Picoïdes Arcticus* from the lower Geyser basin, Wyoming territory, also of the *Goniodes Merriamanus*, the *Tetrao Richardsoni*, and the *Goniodes mephitidis*, found on a *Mephitis* from Fire-Hole Basin, Wyoming territory; of the *Nirmus buteonivorus*, from a *Buteo Swainsonii;* and of *Docophorus Syrnii*, from *Syrnium nebulosum*.

A great number of these insects live between the feathers of birds, and can be more easily observed, since they detach themselves after the death of their host. They are easily found on the skins of birds prepared for museums. These ticks form a family under the name of *Riciniæ*, and this family is divided into two parts, the *Liotheidæ* and the *Philopteridæ*.

Among the many generic divisions, one of the most interesting has received the name of *Trichodectes;* it contains twenty species, one of which lives on the dog, another on the cat, another on the ox; in a word, we discover a distinct species on each of the domestic mammals. It has been said that the *phthiriasis* of the cat is occasioned by the abundance of ricini. The trichodectes of the dog has lately attracted the especial notice of naturalists, and that from the following circumstances.

There is no tape-worm more common in the dog than the *Tænia cucumerina*. But whence comes it ? How is it introduced ? This had been an enigma for many years, at the time when I dissected some dogs infested with *Tænia serrata*, in the Museum of Natural History at Paris. Together with the *Tænia serrata*, the number and age of which I knew beforehand, since I had myself *planted* them, there were found in the intestines of one of the dogs some individuals of the *Tænia cucumerina*. My dogs had taken nothing but milk, and *cysticerci pisiformes*. Were there cysticerci of different kinds in the peritoneum of the rabbit ? The veil is now withdrawn. We have just said that the dog harbours a tick known under the name of Trichodectes, and in this trichodectes lodges the Scolex, we might even say the larva of the *Tænia cucumerina*. Dogs, especially young ones, lick their hair continually, and it is by this operation that the young tænia is introduced. It is by a similar process that the horse introduces the eggs of the Œstrus which are hatched in its stomach.

Many of these ticks live abundantly in birds, and multiply rapidly. The *Liothe pallidum* lives on the·cock, the *Liothe stramineum* on the turkey, the *Philopterus jalciformis* on the peacock, the *Philopterus clavijormis* on the pigeon. It is to be observed that every bird can nourish many different kinds. Fig. 2 represents the tick which infests the sea-eagle, called Pygarg.

Fig. 2—Ricinus of the Pygarg.

Fishes harbour crustaceans instead of ticks, and their number is not less considerable than on mammals and birds. These crustaceans have perplexed naturalists more than once, because they could only regard them as parasites. They live on the produce of cutaneous secretions, and if they improve, as do the ticks, the cleanliness of their host, they are not less useful in a hygienic point of view, for they prevent the accumulation of cutaneous productions.

Among these crustaceans, we must mention the *Caligi* and the *Arguli*, which never become bloated, the *Ancei*, and probably other genera. Instead of the ungainly and unusual forms of true parasites, they all preserve, together with their fishing tackle and locomotive apparatus, their neat and elegant appearance. The sexes even differ only in size. They remain during the whole of their life what they are at the beginning; that is to say, charming in form, with a delicately-shaped corselet, numerous and slender claws, and are as graceful in their movements as when in a state of rest. The greater number of osseous fishes lodge Caligi on the surface of their skin. These fix themselves by means of strong cables, but without sacrificing their liberty. They are usually called fish lice.

Fishermen, when returning from the northern fishery, generally find their vivarium full of these graceful vermin. It may be said that the caligi are common everywhere, and that each species has its own caligi. The fishes of the family Plagiostoma, notwithstanding the hardness of their skin, afford food to some of these; they multiply so rapidly sometimes, that they cover their host as though they took the place of scales. The cod gives lodging to a charming species of a very beautiful shape, which in its turn, affords a resting-place to the Udonella. It is always attached to the ovisacs, and doubtless plays the same part as the Histriobdellæ, so that we shall find the Caligi attending to the toilet of the cod, and the Udonellæ in their turn waiting on the Caligi.

The name Arguli has been given to some crustaceans which resemble the caligi in size and in manner of life, and which principally frequent fresh-water fishes. The *Argulus foliaceus* is the name of the species which has been known for the longest time, and which is most extensively found. It is to be seen on our pikes, carps,

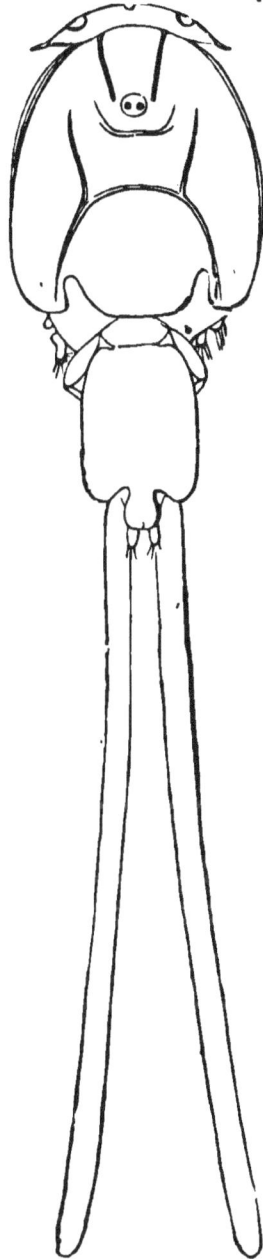

Of the natural size.

Caligulus elegans (fem.)

sticklebacks, and on the greater part of our river fish. Mr. Thorell, in his monograph, mentions twelve species of Arguli proper, and four species of which he composed the genus *Gyropeltis*. Four are found in Europe, two of which are on salt-water, and two on fresh-water fish.

Quite recently, Professor Leydig has made known another species living on the *Phoxinus levis*. Arguli are met with on the fishes of Africa, the Indies, and North and South America. Like the caligi, these animals spontaneously abandon one host, to go and attend to the toilet of another.

Another animal, which has been taken for a Lernæan, deserves to take its place by the side of the Caligi, at least on account of its manner of life. We refer to that singular being which Leydig discovered in 1850 in Italy, while studying the mucous canal of a *Corvina*, at Cagliari, and to which he gave the name of *Sphærosoma*. To judge by the plate and by some details, this *Sphærosoma*, the name of which ought to be changed to *Leydigia*, belongs, if we mistake not, to the same group as the. Histriobdellæ. We are persuaded that the first opportunity will confirm the correctness of this alliance, by the study of its embryonic form. If we had not been able to examine into all the development of the Histriobdellæ, more than one naturalist would have considered them Lernæans, as happened at the congress of German naturalists at Carlsruhe.

If we see many of these crustaceans live a joyous life while young, there are others which seem to practise economy, and to emancipate themselves when they have grown old. Mons. Hesse and Mr. Spence Bate a few years since revealed the secrets of their existence.

Naturalists had recognized some crustaceans under the name of *Ancei*, and others under the name of *Pranizæ*, living together upon fishes, but with very different organs for fishing and swimming. M. Hesse, curious to know the manner of life of the Pranizæ, made observations on them in a small aquarium, and he perceived that the parts of the mouth were all at once transformed into formidable mandibles, which caused them to resemble Ancei. As it had often occurred with respect to other groups, that the same crustacean at different periods of its evolution had been taken for different animals, the naturalist of Brest had some suspicion as to their identity, and soon ascertained by direct observation that he had not been mistaken. The Pranizæ become Ancei, and live upon fishes under their first form, like caligi and arguli. Nothing can be seen which is more curious than these crustaceans, which ride on the back or the sides of fishes, and assume there every possible attitude.

The Pranizæ fix themselves in the mouth and in the gills as well as on the skin. Some are found on sharks as well as on osseous fishes. They fear neither heat nor light, and do very well under damp sea-weed while waiting for the return of the tide. They run and swim with the same facility. When in the condition of Ancei, they lose their agility, and, under this form, all denotes their sedentary habits. They appear to live in holes, at the bottom of which they defend themselves with their powerful mandibles. It has been observed that fecundation is accomplished, as in the *Axolotls*, before the evolution is complete, but that the eggs are not laid until the animal assumes the form of Anceus.

We may here remark that the change of appearance takes place only among the females; the males preserve their dress and their liberty. Some naturalists assert that we must not accept the metamorphosis of either sex as an established fact, except for the purpose of arrangement. All, however, tends to show that Mons. Hesse has fairly interpreted facts; but it appears to us probable that the whole of the history of these strange crustaceans is not fully known.

Fishermen have long since known whale-lice, the *Cyami* of naturalists, of which we have already made mention while speaking of free messmates. They live at liberty on the skin of their host, and multiply with extreme rapidity. These Cyami have a regular form, but completely different from the others, and have given (like the Ricini and the afore-mentioned crustaceans), great trouble to systematic zoologists. The place which they ought to occupy is far from being definitely fixed. At all events they may be considered as a shorter kind of Caprellæ.

As each whale has cirrhipedes which are peculiar to itself, so each has its own cyami. Professor Lütken, of Copenhagen, has made known ten or twelve species, all found on cetacea, in the two hemispheres. The supposed Cyamus, represented by Dr. Monedero as living on the Biscayan whale, is a Pycnogonon.

The Anilocræ and the Nerocilæ, like the Cyami and other genera, establish themselves on the back of a fish which is a good swimmer. Jealous of their liberty, they preserve their oars and their fins, in order to change their convoy, when the desire seizes on them, and do not imitate the Bopyrians, which instal themselves on the narrow branchial cavity of some decapod crustacean, and as soon as they have entered, throw off all their travelling baggage; in fact, there is no other means for them to gain admission; their lot is identified with that of their host; they can no longer live without him. The female only, it is true, thus renounces her liberty; she sacrifices herself, as usual, for her family, while the male, far from giving himself up, preserves his defensive arms, his claws, and his liberty.

The crustaceans called Caprellæ are perhaps not so independent as they appear to be; it is not impossible that their place may be among the crustaceans now under our consideration. They are often found, together with the Tanaïs, on the bodies of cetaceans and chelonians, on plagiostomous fishes, or in the midst of colonies of Sertulariæ. They also establish themselves on buoys when they are well covered with animal life; and we have discovered them in prodigious numbers on a piece of cable which had lain at the bottom of the sea, and the whole surface of which was covered with animals of every kind.

We may here mention the Pycnogonons, the Saphyrinæ, the Peltidiæ, and the Hersiliæ; these crustaceans often crawl over the skins of their congeners, but without ever renouncing their independence; and they are all more or less occupied with the toilet of their neighbours.

We shall place in a second section some animals which have been usually classed among parasites, rather because of their living upon their neighbours than on account of their mode of life. If it is necessary in menageries to have keepers to cleanse the animals themselves, it is as requisite to have others to keep the cages clean, and to remove dung and filth. Many animals perform this office. The rectum of frogs is always literally full of *Opalinæ* which swarm in this cavity, like ants in their ant-hill, and doubtless live on the contents of the intestine.

These Opalinæ are true infusoria, which do not wait till the fecal matters are decomposed, and till the waters are corrupted by their presence; they prevent accidents which might arise, and interfere in time to purify the water from these excretions. There have been found hitherto in the rectum of frogs, and in the different annelids, the Pachydrili, the Clitelides, the Lumbriculi, and the Enchytrei. We have also seen them in the Planaria and the Nemertians. There is no sight more curious for those who are commencing microscopical studies, than the examination of the contents of the rectum of these Batrachians. Van Leeuwenhoeck knew, two hundred years ago, those animalculæ, to which Bloch at a later period gave the name of *Chaos intestinalis*. There are also some Rotatoria, the *Albertiæ* for example, which ought to have a place here, and which Dujardin has described and named. They live in the intestines of the Lumbrici and of snails, and in the larvæ of Ephemerides.

Dujardin first pointed out the *Albertia vermiculus*; since then Mons. Schultze has made known the Albertia of the *Näis littoralis*, and Radkewitz has recognized in the small worm of our gardens the *Enchytreus vermicularis*. Long since, Siebold correctly stated that these animals are not parasites, since they do not live at the expense of their host.

There is a worm in the Philippine Islands, as Professor Semper has informed me, which lodges in the intestines of a fish, with its head usually projecting outwards, and which watches the crustaceans attracted by the excreta of its host; but although it chooses the intestine of its neighbour as a place of shelter, it is not a parasite.

Fishermen affirm, and the examination of the animal's stomach confirms their assertion, that the *Cyclopterus lumpus* feeds on nothing but the excreta of other fishes. Indeed, it is not possible to count the number of intestinal worms known by the name of *Scolex*, which are found in the contents of the stomach and the intestines. Besides this, we have long known the peculiarities of some insects which cannot live except on the dung of certain animals; and there is an example of one of these insects, found in a fossil state, which. anticipated the discovery of the remains of an extinct mammal before unknown in that district. The larvæ of the fly *Scatophaga stercoraria* live only on excrementary matter.

There are also nematode worms which exist under these conditions, and which develop and propagate their species in the intestines as if in the midst of damp earth. The small eel-like creatures so abundant in cow-dung propagate in it; they are not parasites, and are allied to those of which we speak in this chapter.

Besides those attendants which busy themselves about the cleanliness of other animals, we find some whose duties are less extensive, and whose cares are more limited. Many animals produce a greater number of eggs than they can bring to perfection, and those which are decomposed for want of fecundation, or which die in the course of evolution, are under the care of an especial attendant, employed to make away from time to time with the addled eggs, or the embryos that have failed to come to maturity.

In this manner lobsters give lodgings in the midst of their eggs to a worm, which we at first took for a Serpula, and which, after a complete examination, turns out to be one of the Hirudinidæ : we have given it the name of *Histriobdella.* It is as singular in its movements as in its conformation, and its manner of living approaches that of the *Pontobdellæ* of the rays, of which we shall speak subsequently. We announced this discovery a few years ago in the following terms :—

It is known that lobsters, as well as crabs and the greater part of the crustacea, carry their eggs under the abdomen, and that these eggs remain suspended there till the embryos are hatched. In the midst of them lives an animal of extreme agility, which is perhaps the most extraordinary being which has been subjected to the eyes of a zoologist. It may be said, without exaggeration, that it is a biped, or even quadruped, worm. Let us imagine a clown from the circus, with his limbs as far dislocated as possible, we might even say entirely deprived of bones, displaying tricks of strength and activity on a heap of monster cannon balls, which he struggles to surmount; placing one foot formed like an airbladder on one ball, the other foot on another, alternately balancing and extending his body, folding his limbs on each other, or bending his body upwards like a caterpillar of the geometridæ, and we shall then have but an imperfect idea of all the attitudes which it assumes, and which it varies incessantly.

Its rank and its affinities would have given rise to long discussions if we had not made known at the same time its evolution and anatomical structure.

It is neither a parasite nor a messmate; it does not live at the expense of the lobster, but on one of the productions of these crustaceans, much in the same manner as do the Caligi and the Arguli. The lobster gives him a berth, and the passenger feeds himself at the expense of the cargo; that is to say, he eats the eggs and the embryos which die, and the decomposition of which might be fatal to his host and his progeny. These Histriobdellæ have the same duty to perform as vultures and jackals, which clear the plains of carcases. That which causes us to suppose that such is their appropriate office, is that they have an apparatus for the purpose of sucking eggs, and that we have not found in their digestive canal any remains which resemble any true organism. We find the feces, rolled up as balls, placed after each other in their intestines.

The crustaceans also feed other Hirudinidæ. Mons. Leydig has noticed a *Myzobdella* on the *Lupa diacantha*. The fresh-water crab, common in all the rivers o. Europe, nourishes two, the *Astacobdella rœselii*, which lives under the abdomen, or about the eyes, and the *Astacobdella Abildgardi* which especially frequents the branchiæ. Two astacobdellæ on the same crab doubtless play different parts. We should almost venture to assert, *à priori*, that the species in the gills lives as a parasite on the blood of its host, whilst the other, lodged under the abdomen, plays the same part as the histriobdella of the lobster.

We often find among the eggs of the ordinary crab of

our coasts (*Cancer mœnas*) a nemertian which probably performs the same office. He is lodged while young in a kind of firm sheath attached to the abdominal pro-cesses. We have been able easily to study the first phases of its evolution. We have given it the name of *Polia involuta*.

This nemertian had been observed at Messina, and described before by Kölliker under the name of *Nemertes carcinophilus*, and it has just been described and figured anew by Mr. M'Intosh, in a monograph of British an-nelids published by the Ray Society.

The sturgeon seems to give lodging in its eggs to a polyp which plays the same part. In fact, Mons. Ows-jannikoff, at the congress of Russian naturalists at Kiew, described an animal, *Accipenser ruthenus*, which lives in the eggs of the sterlet. Some eggs placed in water for a few hours at first show tentacles on the outside, then a whole colony, and each part consists of four individuals, which have a common digestive cavity, resembling some-what a hydra divided longitudinally in four. Each has six tentacles, two of which are terminated by transparent corpuscles, perhaps nematocysts ; the digestive cavity extends into the arms, as in the hydra ; the mouth is not between the tentacles, but at the opposite pole. They are not all lodged within the eggs ; some are found outside, according to the observations of Mons. Koch. Does not this animal fulfil in the egg of the sterlet, the same office as the histriobdella in the egg of the lobster ?

The eggs of some insects are attacked by very little ichneumons, the *Proctotrupidæ;* they empty them, and then instal themselves in the shell. Mons. Fabre has mentioned, in his memoir on the habits of the *Meloë*, a worm found in an egg.

M. Barthelemy has studied a nematode worm (*Asca-roides limacis*) which inhabits as a parasite the egg of the grey snail; is this not the ordinary worm of the snail which has introduced itself into the eggs?

Many animals establish themselves on their neigh-bours, not to obtain any advantage from them, except to profit by their fins; they are not themselves sufficiently adapted to rapid motion, so they seize a good courser, mount on his back, and ask from him only a resting-place and no provisions. But it is often very difficult to say where commensalism ends and mutualism begins; the cirrhipedes, for example, establish them-selves on a piece of floating wood, or on the bottom of a vessel; on a block of stone, or on one of the piles of a groin; on an immovable animal as well as on a good swimmer.

Some fourteen years ago, Jacobson of Copenhagen wrote an interesting essay, to show that the young bivalves that are found in the branchiæ of anodonts at a certain period of the year are parasitical animals, for which he proposed a new name. But these supposed parasites are only young anodonts, which by the help of a very long cable, which proceeds from their foot like a byssus, attach themselves to their mother, or to a fish which will carry them to a distance.

We see full-grown acephalous molluscs, as mussels and pinnæ, still keep these cables, under the name oi byssus, during their whole life. There are also among distomians, worms which though they are hermaphro-dite, couple two and two, and have this additional peculiarity, that while one increases rapidly the other becomes atrophied.

An Egyptian distome, which lives in man, gives an instance of this peculiarity, as well as the *D. filicolle,* which inhabits a fish (*Brama Raii*). The caligi which live on the skin of fishes are, when young, fastened by a cord which comes from the anterior edge of their carapace : while quite little, they put themselves under the protection of a kind neighbour, and allow themselves to be led by him.

The new tubularia, which we have dedicated to our learned colleague Dumortier, often fixes itself on the carapace of ordinary crabs, and causes itself to be conveyed like the Echeneis; the tubulary observed by Gwyn Jeffreys, close by the eye of the *Rossia papillifera,* a cephalopod mollusc, perhaps belongs to the same species.

Every colony of campanulariæ or sertulariæ lodges a crowd of messmates and mutualists ; and there are a great number of crustaceans and polyps of all sizes which serve as an abode for infusoria of every kind. Some establish themselves on the carapace or on the swimming appendages, as in a carriage ; others on one of the gills, which renders their mode of life more easy, and the danger less great. An amphipod very extensively spread over our sea-coasts, the *Gammarus marinus,* usually has its appendages covered with *Vayinicola crystallina.*

CHAPTER V.

PARASITES.

"En plongeant si bas dans la vie, je croyais y rencontrer les *fatalités physiques*, et j'y trouve la justice, l'immortalité, l'espérance."—MICHELET, *l'Insecte.*

THE parasite is he whose profession it is to live at the expense of his neighbour, and whose only employment consists in taking advantage of him, but prudently, so as not to endanger his life. He is a pauper who needs help, lest he should die on the public highway, but who practises the precept—not to kill the fowl in order to get the eggs. It is at once seen that he is essentially different from the messmate who is simply a companion at table. The beast of prey kills its victim in order to feed upon his flesh, the parasite does not kill; on the contrary he profits by all the advantages enjoyed by the host on whom he thrusts his presence.

The limits which separate the animals of prey·from the parasite are usually very clearly marked; yet the larva of the ichneumon, which eats its nurse, piece after piece, resembles a carnivorous animal as much as a parasite. There are indeed certain animals which take advantage of the good condition of their Amphitryon, but which render to him in return precious services. Thus those which live on the produce of the secretions, or which clear the system of useless materials in exchange for the hospitality which they receive, are not true parasites. These services are of a very different character, and the duties which they sometimes perform for each other are in some respects analogous to medical care.

Every animal has its own parasites, which always come from without. With some few exceptions, they are introduced by means of food or drink. In order to ascertain their origin, the naturalist must beforehand study the food, that is to say, the prey or the plant which furnishes the habitual nourishment of the host which gives them shelter.

A carnivorous animal, however, does not in general content himself with a single kind of prey—one voracious animal of this class devours all that comes in its way; another, more of an epicure than a glutton, chooses with more discernment. But in the midst of this varied kind of food there is always some species which forms the staple of the usual bill of fare, and it is necessary to find out what this is if we wish to ascertain the parentage and the metamorphoses of the parasite, since it is that which conducts the parasite to its new destination. The mouse is destined to the cat, and the rabbit to the dog; in the same manner, each one of the herbivora is intended to be the prey of a carnivorous animal, if not larger and stronger than itself, at least more cunning. It is of great importance to discover the animal which conducts the new-comer into his habitation. When we know it, we have only to introduce into it the stranger guest, that sooner or later he may pass into the body of his accustomed Amphitryon. In order thoroughly to know these sedentary or vagabond populations, we must not only study them at the different periods of the year, and under all the conditions of their irregular life, but it is necessary to follow them from the moment that they quit the egg till their complete evolution, closely noticing all that relates to their reproduction.

In the dung of the cow, by the side of the elegant *Pilobolus*, live masses of small eels, born in the stomach of the animal, which wind and twist like microscopical serpents, and do not seek the slightest help from the organ which shelters them. They are hatched in the interior of the stomach, as if it took place in the meadow. These little eels have evidently only the appearance of parasites, and it may be that they render some service in some of the organs through which they pass. This may also be the case with those which live on the feces of others, or which, lodged in the rectum, watch for the prey which is attracted by the odour. These, especially the latter, are rather messmates than parasites. True parasites are animals entirely dependent on their neighbours, unable to provide for themselves, fed entirely at the expense of others. It is generally supposed that parasites are exceptional beings, requiring a place by themselves in the animal hierarchy, and knowing nothing of the world except the organ which shelters them. This is an error. There are few animals, however sedentary they may be, which are not wanderers at some period of their lives, and it is not even uncommon to find some which live alternately as noblemen or as beggars. Many of them only deserve to be placed among paupers when they are in their infancy or at the approach of adult age, for they only seek for help at the beginning or towards the end of their career. These are very numerous, and more than one species change their dress so completely that they can no longer be recognized. Finding with their host both food and lodging, they throw off their fishing and travelling gear, settle themselves comfortably in the organs which they have chosen, and having got rid of the baggage which connected them with the outer world, preserve only their sexual organs.

As to the rank which these parasites occupy in the scale of being, it may be said that there is no especial class of parasites; and worms are not distinguished in this respect, except by having a greater number of species subject to this rule. All classes among invertebrate animals include parasites.

It is also an error to suppose that the whole species, the young as well as the old, the males as well as the females, are always parasites; often the female, not being able to provide for the necessities of life, seeks for food and shelter, while the male continues his nomad life. Therefore the female alone puts on the pauper's dress, and by a recurrent development, assumes sometimes such a singular appearance that the male no longer resembles her. One cannot say that the females constitute the *beau sexe* in this group, since they are often so monstrous in form and size that their appearance has nothing in common with a perfect animal; their body is deprived of all its exterior organs, and there often remains only a skin in the form of a leather bag, without any distinguishing character.

What is still more astonishing, is to meet with males which, under the conditions to which we have just alluded, come at last to seek for assistance from their own female, so that she has to provide for all; and the charitable animal which comes to her help takes the whole family under his charge. Assistance is thus thoroughly organized in the lower world; neighbours are found which serve as a *crèche* for the indigent when they first quit the egg, others as a hospital for the infirm adults or the females, and others again play the part of innkeepers for all, instead of affording a place of refuge for some privileged individuals.

There are but few animals, if indeed there are any, which have not their peculiar parasites. Of all the fishes of our coasts we have never found but one which had none; and perhaps, could we observe this fish in different latitudes, we might find that it had its poor dependants as well as the rest.

Thus we may assume that no animal is free in this respect, and man himself regularly affords hospitality to many of them. We feed some with our blood and our flesh; there are some which lodge on the surface of our skin, others in the interior of our organs; some prefer to establish themselves on children, others on adults. The name alone of some is sufficient to make us shudder, while others live peaceably in some crypt, without our suspecting their presence. Who is there that does not nourish some acari, of the genus *Simonea*, in the membrane of the nose? In fact, man gives a home to some dozens of parasites, and the presence of the most terrible among them constitutes, in certain countries, a condition of health which is envied. The Abyssinians do not consider themselves in good health, except when they nourish one or many tape-worms.

Among the animals to which man gives his involuntary assistance, we may mention first, four different Cestoidea, or tape-worms, which live in the intestines; three or four Distoma, which lodge in the liver, the intestines, or the blood; nine or ten Nematodes, which inhabit the digestive passages or the flesh. There are also some young Cestodes, named *Cysticerci, Echinococci, Hydatids*, or *Acephalocysts*, which find in him a *crèche* to shelter them during their life. These always choose enclosed organs, like the eye-ball, the lobes of the brain, the heart, or the connective tissue. We also provide a living for three or four kinds of lice, for a bug, for a flea, and two ascarides, without mentioning certain inferior organisms which lurk in the tartar of the teeth, or in the secretions of the mucous membrane.

There are some animals which harbour few inhabitants, while there are others that keep up a great retinue; and it is not always, as we have already said, that those who give lodging to but few enjoy the most excellent health. We might give as an instance of this, a fish which is known to all, the turbot, which as well as the woodcock is highly prized, though both have their intestines literally obstructed by tape-worms and their eggs. We have never opened one, large or small, lean or fat, which had not its intestines filled with cestode worms. They are so numerous as to form a kind of cork, which one might think intended to close the passage of the pylorus.

Some authors give remarkable instances of the abundance of parasites. Nathusius speaks of a black stork, which lodged twenty-four *Filariæ lobatæ* in its lungs, sixteen *Syngami tracheales* in the tracheal artery, besides more than a hundred *Spiropteræ alatæ* within the membranes of the stomach, several hundreds of the *Holostomum excavatum* in the smaller intestine, a hundred of the *Distoma ferox* in the large intestine, twenty-two of the *Distoma hians* in the œsophagus, and a *Distoma echinatum* in the small intestine. In spite of this affluence of lodgers the bird did not appear to be in the least inconvenienced.

Krause, of Belgrade, mentions a horse two years old, which contained more than five hundred *Ascarides megalocephalæ*, one hundred and ninety *Oxyures curvulæ*, two hundred and fourteen *Strongyli armati*, several millions of *Strongyli tetracanthi*, sixty-nine *Tæniæ perfoliatæ*, two hundred and eighty-seven *Filariæ papillosæ*, and six *Cysticerci*. When we consider how many eggs a single worm produces, we can understand how it is that so few animals escape being invaded by them.

Sixty millions of eggs have been counted in a single nematode, and in a single tape-worm, or rather in a colony, even a thousand millions of eggs. Even the very animals which live as parasites, harbour others in their turn. We find parasites on parasites, as we find messmates upon messmates. Almost all writers on this subject give examples of these; some in the larvæ of ichneumons, others in the lernæans, and we have more than once met with nematodes in different crustacea still attached to their host.

In order to understand thoroughly the living furniture of an animal, especially of a fish, it is necessary to examine it while young; the feces are the *Kitchen-mid-*

6

dings of the stomach ; it is from them that we can appre-
ciate the bill of fare of each.   This study of the food will
one day excite much interest, not only in a scientific
point of view, but also with reference to fishing as an
occupation.

There are some animals which are infested at every
period of their life, and at every season ; others in far
greater number only during their youth, and they gather
in at the commencement of their life the harvest for the
rest of their days.   The greater part of parasites, especially of fish, are introduced with the first nourishment.
As soon as they issue from the egg, young rays, like
young turbots, are already stuffed with worms which
afterward obstruct the digestive organs.   The stomach of
each of these fishes is like a filter which allows every
thing which is food to pass, but detains on its passage
and without any change all that is living.   When we
examine the stomach and observe the food in its different
degrees of digestion, we see distinctly the worms coming
out of their holes, wallowing in that which physiologists
call chyle, and choosing afterwards at their convenience
the place where they may completely develop themselves.
At the end of a few days, the fish may have swallowed an
innumerable quantity of small animals, and if each of
them introduces some worms, we can easily understand in
how short a time the intestine becomes literally filled.

There is no organ which is sheltered from the invasion of parasites : neither the brain, the ear, the eye,
the heart, the blood, the lungs, the spinal marrow, the
nerves, the muscles, or even the bones.   Cysticerci have
been found in the interior of the lobes of the brain, in
the eye-ball, in the heart, and in the substance of the

bones, as well as in the spinal marrow. Each kind of
worm has also its favourite place, and if it has not the
chance of getting there, in order to undergo its changes,
it will perish rather than emigrate to a situation which
is not peculiar to it. One kind of worm inhabits the
digestive passages, some at the entrance, others at the
place of exit; another occupies the fossæ of the nose;
a third the liver, or the kidneys.

We may even divide parasites into two great categories, according to the organs which they choose:
those which inhabit a temporary host, almost always
instal themselves in a closed organ—in the muscles, the
heart, or the lobes of the brain; those, on the contrary,
which have arrived at their destination, and which,
unlike the preceding, have a family, occupy the
stomach with its dependencies, the digestive passages,
the lungs, the nasal fossæ, the kidneys, in a word,
all the organs which are in direct communication with
the exterior, in order to leave a place of issue for their
progeny. The young ones are never enclosed. Even
the blood is not free from these animals, but there
are few which lodge there, except during the act of
migration.

In Egypt, Dr. Bilharz observed a distome in the
blood of a man (*Distoma hæmatobium*); the *Strongylus*
of the horse has been long known, which causes serious
injuries in its vessels (*Strongylus armatus*); as also the
strongylus of the dolphin and of the porpoise (*Strongylus
inflexus*), and the filaria of the dog (*Filaria papillosa*);
and some are also found in the blood of many birds,
of reptiles, batrachians, and fishes; so that there is no
class of vertebrates which escapes.

There are some which, like leeches, seek assistance
from their neighbours, but are content to snatch their
food as they pass, and only attach themselves for a
short time to the host which they despoil; they retain
their fishing or hunting tackle, as well as their organs of
locomotion.   These parasites, which never take up their
lodging on the host which nourishes them, have no
sooner sucked his blood, or devoured his flesh, than they
resume their independent life.

They do not disfigure themselves, nor put on any
special costume, like those which seek a permanent
abode.   Gluttony is not with them the only moving
principle of existence; they do not forget what they owe
to the world, and keep up an appearance which allows
them at all times to present themselves afresh.

Parasites are scattered over every region of the
globe; they choose their place, and observe, like all
living creatures, the laws of geographical distribution.
All do not inhabit the animal kingdom; some seek for
assistance in vegetable life.   Many insects lay their eggs
in seeds or fruits, and their progeny, as soon as they
are hatched, find abundant nourishment in the sap or
in the farina stored up for the young plant; others
pass into a state of lethargy while the seed is dry,
and recover their activity every time that they receive a
little humidity.

The female of a coleopterous insect deposits its eggs
in the nut, and in proportion as this grows, the young
larva devours the kernel.   When it is brought to table,
it encloses only the skin and the excretions of the larva.
A weevil establishes itself in a similar manner in cereal
plants, and, small as it is, it may produce great calamity

by multiplying in granaries. There are even worms which lodge in certain of the graminaceæ, and get completely dry with the envelope which contains them, without ceasing to live. Their life is suspended till the day when the seed is sufficiently softened in the earth or the water.

We have seen that each parasite has its host: we must have a particular name to designate it. But that does not imply that if it find not its dwelling-place it must perish. It may only live some time at the expense of its neighbour, and thus pass for its parasite. Naturalists are occasionally deceived. Thus, they once believed in the passage of the Schistocephalus of the stickleback into the intestines of certain birds which eat them, and in which they are only found accidentally. The Ligulæ of the Cyprinidæ, found in the intestines of the cormorant or the goosander, are not, in our opinion at least, worms peculiar to these birds. They are strangers which must either emigrate again or die. Acari which originally belonged to mammals and birds, have been found living on man, causing prurigo, or even serious maladies, and yet these parasites are not regarded as peculiar to our species. We might cite other examples. Who has not been annoyed by the flea, which abandons for an instant the dog, its natural host?

Among these free parasites, many do not attach themselves to a particular species, and well deserve the title of cosmopolitan parasites. Thus we see that the *Ascaris lumbricoides,* so common among children, lodges also in the ox, or the horse, the ass, and the pig. The *Distoma hepaticum,* which is a parasite peculiar to the sheep, if we may judge by its abundance in this animal, may find its way into the liver of man, or into that of the hare, the rabbit, the horse, the squirrel, the ass, the pig, the ox, the stag, the roebuck, and different species of antelope. It is to be remarked that all these animals have a vegetable regimen. By drinking the water which contains the cercaria of this species, they grow infested by this singular lodger. The large Echinorhyncus (*E. Gigas*) has been found in the dog, and the pig, perhaps in the phocinæ; and instances are mentioned in which it has even migrated into man. The *Gordius aquaticus* appears to live and develop itself in different species of insects; and among the articulated parasites, we meet with the *Ixodes ricinus*, commonly called the tick, on the dog, the sheep, the roebuck, and the hedgehog; and instances are given of its presence on man. It has been long since proved in menageries and zoological gardens, that the *Acarus* of the camel is able to give a cutaneous disease to man.

As we have before said, there are many parasites which require to be studied in order to determine the host peculiar to each of them; although parasites sometimes lose their way, and introduce themselves into the wrong neighbour, yet they can live there but a short time. Instances have been known, in which the larvæ of flies have penetrated into man accidentally by the mouth or the nostrils. Reptiles have been known to live a certain time in the stomach. A German physiologist, Berthold, professor at the University of Göttingen, has given an account of all those which have been found under such circumstances, and the number of them is considerable; he has written a memoir on the abode of living reptiles in man.

Among other instances, this naturalist mentions the case of a boy of twelve years of age, who, in 1699, after suffering acute pain, voided from the intestines nearly one hundred and sixty four millipedes, four scolopendræ, two living butterflies, two worm-like ants, thirty-two brown caterpillars of different sizes, and a coleopterous insect. These animals lived from three to twelve days. This is not all: the same child, two months afterwards, voided four frogs, then several toads, and twenty-one lizards, and sometimes a live serpent was seen for a moment at the bottom of his mouth. Happily for science, we do not see such things seriously related in books at the present day.

The size of parasites is very various: Boerhaave mentions a bothriocephalus three hundred ells in length; at the Academy of Copenhagen, it was reported that a solitary tape-worm (*Tænia solium*) had been found eight hundred ells long. Female strongyli have been seen from two decimètres to one metre in length; and *Gordii* of two hundred and seventy millimètres. We have found in a fish a worm which lived rolled up like a ball, and which measured, when unrolled, more than a mètre.

Parasites present an extraordinary variety of forms, and the differences between the sexes in size as well as in appearance are greater than in any other group of animals. The male of the *Uropitrus paradoxus;* the Urubu of Brazil, has the usual form of a round long worm, while the female resembles a ball of cotton, without the slightest analogy with the other worms of the order. The Lernæans also have females excessively various in size and appearance, while the males generally resemble

each other in their external characters. What is not
less remarkable is, that hermaphrodite worms often
unite in couples, and that only one of the two seems to
perform the function of a female, and increases in size
(*Distoma Okenii, Bilhartzia*). It even happens that the
union is so complete that the species appears formed of
two individuals fastened to each other. The Diplozoa
show us a curious example of this. The gills of breams
are usually infested by these last-mentioned worms.
Nothing is more strange than to see all these individuals
united two and two as if soldered together, each pre-
serving its mouth and digestive canal, and producing
eggs which give birth to isolated individuals. We sometimes see males so completely absorbed in their females,
even in an anatomical point of view, that they only
represent a fragmentary apparatus. The male of the
Syngami is so obliterated, that when compared with the
other males of its order it is only a testicle living on
the female.

Should an organ infested with worms be considered
diseased, simply on account of their presence ? We hesitate not to say that, as long as these guests cause no
disorders, there is no pathological condition. The child
which has *Ascarides lumbricoides* in its stomach is not
necessarily ill. All animals in a wild state always
have their parasites ; they lose them rapidly when in
captivity.

The Abyssinians do not take medicine when they
have tæniæ ; on the contrary they are in a better state
of health. Do we not find medical men prescribing the
employment of leeches, and consequently calling in the
assistance of certain parasitical animals ? This action,

far from being a cause of sickness, is in this instance
a remedy, and no one can foresee all that science has
a right to expect from the salutary effects of certain ,
parasitical worms on the system. There are, if we
mistake not, many discoveries in store for observers
in this order of investigation.

But here, as in all things, excess is hurtful. Certain
organisms, developing themselves immoderately, may
break the harmony necessary between the parasites
and the host which they frequent. It has been found
recently that many morbid affections, as the potato
and vine diseases, have for their origin only the
abnormal development of certain microscopic beings
hidden in the organism.

It is found, that in Egypt, a distoma is developed
in the blood, and occasions a very severe malady,
scarcely known to physicians. In Iceland, a cestode
causes the death of a third part of the population.
Worms develop themselves in the eye, and may even
cause blindness; the *Cœnurus* of the sheep causes giddi-
ness, and becomes fatal to the animal which harbours it.
The chlorosis observed in Egypt and Brazil must, it
appears, be attributed to a considerable development of
a nematode worm, which lives in the small intestines,
and which naturalists know under the name of *Dochmius
duodenalis;* and lately the Trichinæ set all Europe in a
state of excitement, and trichinosis was for a time more
dreaded than cholera. In spite of all these accidental
circumstances we think that the animal which possesses
its ordinary parasites, far from being ill, is in a normal
physiological condition.

When we consider these animal parasites in general,

one would think that their tenacity of life is very feeble, and that the slightest derangement would be sufficient to kill them. It is not so; on the contrary, some of them can be entirely dried up, and return to life every time that they are moistened; and the eggs of some of them resist the most violent reagents. We have known eggs preserved for years in alcohol, in chromic acid, and in other agents which destroy life everywhere else; and then give birth to embryos directly they are placed in pure water or damp earth.

Some years ago they had no idea of the migration of animals from one body to another. As we have said elsewhere, Abildgard, half a century ago, made experiments on the worms of fishes which he caused ducks to swallow, but these experiments had no result, and formed rather an obstacle to ulterior progress, than an approach to truth. The worms of fishes have been known to live in birds; but these worms were only there as adventitious parasites. Liguli live some days in the goosander, but they do not maintain their position.

Our great initiator into the world of parasites, Mons. Siebold, arrived also at a conclusion which could not be maintained. Having observed, with his habitual sagacity, that the cysticercus of the mouse is the same worm which lives in the cat, he published his opinion that the eggs of this tænia had lost their way in the mouse, that the young worms had become sick there, and that in the cat alone, they could be healthily and completely developed. It was like a plant lost on a soil where it could not live, and still less flourish. May I be permitted to state by what means we have arrived at the knowledge of the transmigration of worms?

I had commenced the study of encysted Tetrarhynchi in the peritoneum of the Gadidæ in 1837. Ten years afterwards, shortly after a visit from my learned friend, Mons. Kölliker, I discovered that this world of parasites did not live such a monotonous life as was supposed. I ascertained by my dissections of fishes, that the tetrarhynchi also, which were supposed to be disinherited by Nature, knew how to vary their pleasures; that instead of spending their whole life in a prison cell, they change their home at a certain age, and pass the latter part of their existence in more spacious habitations.

I had seen the *Tetrarhynchus agamus* inhabiting a cyst in the peritoneum of the gadidæ, and I had met with the same tetrarhynchus completely developed and sexual in the spiral intestine of the voracious fishes known under the name of squalidæ, or sharks. This caused me to write to the Academy of Brussels, at the meeting on January the 13th, 1849, that the order of vesicular worms, admitted by all helminthologists, ought to be suppressed.

These worms began to be understood when these cysticerci ceased to be regarded as sick creatures. Siebold had mistaken the *crèche* for the hospital, and instead of seeing in the cysticercus a young animal full of life and of the future, he looked upon it as a gouty individual, ready to breathe its last sigh.

These fish had directed me in the right road; I had closely followed up certain very characteristic worms, which lived under a very simple form in certain fishes, and which, passing with their host into the stomach of another, finished in the latter their toilet and their evolution. I had been a witness of all their changes of form from the cradle to the tomb, by following them from fish to fish, or rather from stomach to stomach. In fact these parasites are perpetually on their journey, and constantly changing their host, and at the same time their dress and mode of locomotion, so that frequently, at the end of their voyage, they preserve only shapeless rags to cover their eggs or their offspring.

That which adds still more to the difficulty of recognizing them is, that while young they are often enveloped in swaddling clothes which nevertheless permit them to wander freely; then in a simple robe, in keeping with the home which shelters them; and at last in a wedding dress, which hides the eggs and the apparatus which produces them. The nymph in her virgin condition has none of the attributes of future maternity.

It is in this category that we find the Distomes, so common in all the classes of the animal kingdom. This is not all: frequently, among these various forms, these animals when young produce little ones, which in no respects resemble the others, and are not even formed in the same manner. As soon as they quit their swaddling-clothes, they increase by gemmation, and without sexual union, while those which are produced from buds increase sexually. Thus the daughter does not resemble her mother, but her grandmother. This phenomenon has been known by the name of alternate generation; we have called it *digenesis*.

But all parasites do not resemble those distomes, which change several times both their host and their costume. We find some of them, which the mother deposits with care in the body of a neighbour, and which pass all their early life in the viscera of an alien mother.

Such are the Ichneumons, beautiful winged insects, which perfidiously insert their eggs in the body of a living caterpillar, whose internal part serves at the same time for a cradle and for food. The young larva devours organ after organ, beginning with the least important, till the last serves for the formation of the last members of the winged insect.

More unfortunate are those which are kept under the bolts and bars of their host from their early youth to mature age; they have no participation in the great banquet of life, except it be in the pleasures of the table and of love. We also find some parasites which occupy different organs in the same animal, and which have different sexual attributes according to the situation which they inhabit. We know some which are hermaphrodite in the rectum or in damp earth, and whose young ones, having the sexes separate, live as parasites in the lungs.

Parasites are not usually reproductive in the animal which they inhabit. They respect the hearth which shelters them, and their progeny are not developed by their side. The eggs are expelled with the feces, and sown at a distance for other hosts.

Parasites may be divided into several categories. We may bring together in the first of these, a certain number of animals, which, without being true parasites, seek for a place of shelter, and, either on account of their wretchedness or their misery, require this protection in order that they may live.

In the second category, we may place those which live at complete liberty, and only require for their sustenance the superfluities of their neighbours; they take great care of the skin of their host, and use it sparingly. Some also are found which cannot live without assistance, but repay it with some service. Often, indeed, they associate with their host, and live on a footing of perfect equality with him; and besides these are found associations in which equality is by no means recognized, and where labourers or even slaves perform the work disdained by their masters.

In the last category we shall arrange true parasites, which take both their lodging and their food. And here, again, we shall meet with three distinct subdivisions.

The first includes those which travel from one hotel to another before they arrive at their destination; to-day they lodge in a prawn, to-morrow in a gudgeon, then in some fish which preys upon others, as the perch or the pike. These are nomadic parasites, which do not stop or think of family life until they have found the hotel for which they are destined.

Sometimes the parasite gets into a wrong train, and not being able to retrace his steps, he remains at a station where no other train will take him up. He is condemned to die in a waiting-room.

In the last subdivision, we have parasites that have arrived at their destination, occupying themselves in future only with the joys of a family.

Thus we find some which are really at home, and others which are on their journey, sometimes on the right road, and at others, wandering and lost in an alien " host." The former are *autochthonic* parasites, the others are foreigners. We may say that each animal species has its proper parasites, which can live only in animals which have at least more or less affinity with their pecular host. Thus the *Ascaris mystax*, the guest of the domestic cat, lives in different species of *Felis*, while the fox, so nearly resembling in appearance the wolf and the dog, never entertains the *Tænia serrata*, so common in the latter animal.

The same host does not always harbour the same worms in the different regions of the globe which it inhabits. This relates both to the parasites of man, and to those of the domestic animals. Thus the large tape-worm of man, which naturalists call *Bothriocephalus*, is found only in Russia, Poland, and Switzerland. A small tape-worm, *Tænia nana*, is observed nowhere except in Abyssinia; the *Anchylostoma* is known at present only in the south of Europe and the north of Africa; the *Filaria* of Medina, in the west and the east of Africa; the *Bilharzia*, that terrible worm, has only been found in Egypt.

There are also parasitic insects dreaded by man, as the *Chigoe* (*Pulex penetrans*) which, happily, is only known in certain countries. Some, however, have become cosmopolitan, since man has introduced them wherever he has established himself.

The mammalia which live on vegetable diet have Tænia without any crown of hooks, and man, according to his teeth, ought only to nourish the *Tænia medio-canellata*. We find in a work on the Algerian Tænia, by Dr. Cauvet, that it is the *Tænia inermis*, that is to say, without hooks, which is the species common in Algeria. Among fourteen tæniæ which he had occasion to examine, there was not a single *Tænia solium*. I have said long since, that this species ought to be less widely spread than the tænia without hooks. The *Tænia solium* comes from the cysticercus of the pig, the other from that of the ox; and Dr. Cauvet has ascertained that the latter, in the state of cysticercus, has already lost its crown.

We find extinct fossil genera and species in all the classes of the organic world. Is it the same with worms and animals of other classes which are only known in the condition of parasites? Had the Ichthyosauri and the Plesiosauri worms in their spiral cœcum like plagiostomous fishes, which resemble them so much in the digestive tube? We do not doubt this, and we should have been glad to give some demonstration of it. For this purpose, we have made a collection of the coprolites of these animals, but we have not yet succeeded in getting slices thin enough or sufficiently transparent to discover the eggs or the hooks of their cestode worms.

Not long ago, the partisans of spontaneous generation found in the class of worms their principal argument for their old hypothesis, and it was even after the publication of my treatise on intestinal worms that this question, which seemed forgotten, was taken up again by Pouchet. At present, they appear to have given up parasites, which reproduce their kind like other animals, and to have fallen back upon the infusoria, the last intrenchment which remained to the partisans of spontaneous generation, whence Mons. Pasteur has scientifically dislodged them. It is evident to all those who place facts above hypotheses and prejudices, that ᐟspontaneous generation, as well as the transformation of species, does not exist, at least, if we only consider the present epoch. We are leaving the domain of science if we take our arms from anterior epochs. We cannot accept anything as a fact, which is not capable of proof.

CHAPTER VI.

This first category of parasites includes all those which are not enclosed, and which live at the expense of others, without losing the attributes and advantages of a wandering life; they are as free as the vulture or the falcon which pursues its prey. We shall not, however, include among them the parasitical kite of Daudin, which tears from the hands of the traveller a piece of the flesh which he is preparing in the open air, nor the small Egyptian plover, which keeps the teeth of the crocodile clean. The former is a pirate, a highway robber; the plover, on the contrary, is a kind neighbour, an attendant who performs valuable services.

We are more correct in considering as parasites the Vampires (*Phyllostoma*), those audacious bats of South America, which settle on the sleeping traveller or his beasts, and suck their blood by means of the sharp papillæ of their tongue. These animals are winged leeches which bleed their victim and pass on. We place among free parasites the greater part of leeches, some insects, and a certain number of arachnida, crustaceans, and infusoria.

As we have mentioned free messmates, so we have

free parasites, which take advantage of their host, but with prudence and economy ; they ask from him nothing but his blood, and sometimes render him important services. Many of these animals, both messmates and parasites, have at present been only provisionally classified, and cannot be definitely arranged till more observations have been made. It is not always so easy as it may be thought to determine exactly the relations which certain animals have with each other. We must pry very narrowly before we can ascertain the motives which act on this inferior order of beings. It is among free parasites that we find those organisms which are generally called vermin, and which seem the more capable of injuring their neighbours since they can the more easily escape detection. These creatures, though they are called vermin, excite no more repugnance in the mind of the naturalist than the other works of creation; and St. Augustine did not exclude them from his thoughts when he exclaimed, *"Magnus in magnis, maximus in minimis."*

Leeches drink the blood of their victim, and when they are gorged to the very lips, they fall off, taking a siesta for weeks or months. Thus enjoying a repast at very long intervals, it is useless for them to continue longer at table ; and this is therefore another reason that they should usually preserve their organs of locomotion, that they may use them after their long period of digestion.

Like the annelids, they do not change their form, and as they are only attached to their host for a short time, naturalists have not thought fit to place them among parasitical worms, or Helmintha. However, if we pass from the higher kind of leeches to those which live at the expense of fishes, of crustaceans, and especially of molluscs, we see that the desire of possessing a lodging is developed by insensible degrees, and that the lower kinds, are by their form, their organization, and their mode of life, as dependant as the greater part of the helmintha. Thus we see Hirudinidæ on the Mya, an acephalous mollusc, incapable of quitting their place, firmly fixed on the walls of the stomach of their host, and living quietly at his expense. They are called *Malacobdellæ*, and they have been so ill-treated by Nature, that it is necessary to submit them to minute investigation in order to determine their parentage.

The most well-known leeches are those which attack man and the other mammalia, but some are also found on other vertebrate animals, especially on fishes. Their organization is always proportioned to that of the host which they frequent; thus, the simpler their host, the lower is their organization. The mollusc harbours hirudinidæ much lower in the scale than those which are found in fishes, and especially in mammals.

Vampires make use of the papillæ of the tongue, and also of their teeth, which act as so many lancets; leeches apply their toothed lip, saw asunder the epidermis, and with the mouth applied to a network of capillary vessels, suck till they fall off, intoxicated with blood.

We give here the different appearances which the skin assumes after the bite of a leech. (Fig. 4.)

Fig. 5 (1 and 2) represents the jaws; 1, the jaws in their usual position; 2, a single jaw, to show its outer edge, which is cut with teeth like a saw.

Fig. 6 shows a leech with a section of its digestive

tube. The letters *d d* indicate the different cavities of
the stomach, which are filled in succession. We see in

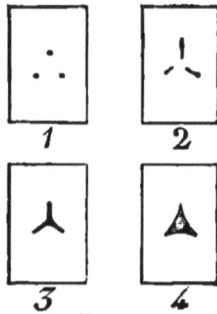

Fig. 4.

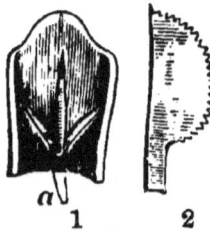

Fig. 5.

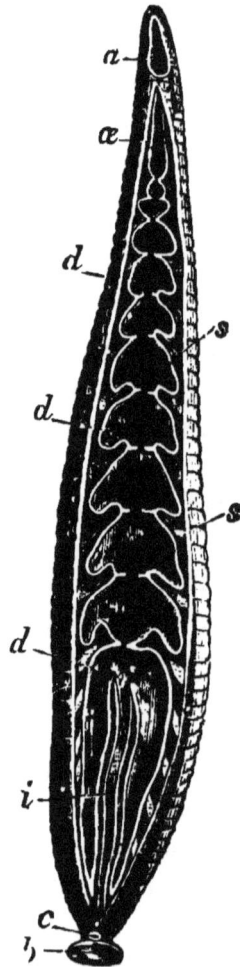

Fig. 6.

Fig. 4.—Different forms of the bite of a Leech.

Fig. 5.—1. Sucker, open; *a.* jaws. 2. One of the jaws magnified.

Fig. 6.—Section of a Leech. *a.* anterior sucker; *b.* posterior sucker; *c.* anus; *d* stomach; *œ.* œsophagus; *i.* intestine; *s.* glands of the skin.

the fore part, the anterior sucker with the mouth, and
behind, the posterior sucker with the anus. At the

side of the stomach are seen traces of the glands of the skin.

We find a great variety in the mode of life of these hirudinidæ; and if we sometimes meet with some which are sober and delicate, the greater part show a voracity of which it is difficult to form any idea. A leech has been met with in Senegal which draws a quantity of blood equal to the weight of its body. There are leeches which devour entire earth-worms. Fortunately the greater species are not the most voracious: we might feel rather uneasy in the midst of leeches similar to that which Blainville has described under the name of *Ponto-bdella lævis*, which is not less than a foot and a half in length.

It is generally thought that all leeches are aquatic, but this is a mistake. In the warm regions of the Old and New World, there live in the midst of the brush-wood, leeches which attack the traveller as well as his horse, and suck the blood of both without their perceiving it.

Hoffmeister gives the following account with reference to small leeches in the island of Ceylon:—

He had amused himself one evening by collecting some phosphorescent insects which were hovering around him in considerable numbers; on entering afterwards a lighted room, he perceived streaks of blood all down his legs. This was the effect of the bites of leeches. These creatures, said he, made a painful impression on me, the remembrance of which was terrible. This same leech, which bears the name of *Hirudo tagalla*, or *Ceylonica*, lives in the thickets and woods of the Philippine Islands. There also it attacks horses as well as men. It has also been noticed on the chain of the Himalayas, 11,000 feet above the level of the sea. Japan and Chili also have terrestrial leeches. The *Cylicobdella lumbricoides* is a blind leech, which has been found by F. Müller in damp earth, in Brazil.

The aquatic leeches are better known, and with but few exceptions, the accidents produced by them are little to be feared. In Algeria it is not uncommon, as army surgeons tell us, to see soldiers, while drinking spring water, swallow small leeches which may do them injury.

We find from official reports that the French soldiers often suffered, during the campaigns in Egypt and Algeria, from an aquatic leech (*Hœmopis vorax*), which attacked the mouth and the nostrils, and did not respect man any more than horses, camels, and oxen. The leech discovered by Dr. Guyon under the eyelids and in the nasal fossæ of the crab-eating heron of Martinique, is probably a monostomum, and not one of the hirudinidæ. Leeches have also been found on turtles under the name of *Eubranchella Branchiata*. Say saw one on a chelonian, and others on tritons and frogs.

It is especially upon fish that these worms are found, and we cannot hesitate to consider the greater part of them as true parasites. We have described a whole series of them which live upon marine fishes, especially on the barbel, the bass or sea-wolf, the halibut, the dab, and different species of gadidæ. A. E. Verril published last year the description of several kinds of American leeches, among which we see two which infest a fish (*Fundulus pisculentus*) of West River, near Newhaven. A large and beautiful species, which is known by the name of *Pontobdella*, is also found upon the Rays.

A very skilful naturalist, Mons. Vaillant, has lately
made these animals the subject of study.   Mr. Baird, in
1869, made known four new Pontobdellæ, one from the
coast of Africa, two from the straits of Magellan, and
one from Australia, found in one of the Rhinobatidæ.
But the most interesting in every point of view are the
Branchellions, which inhabit the electrical fishes known
under the name of torpedoes, and which do not fear to
choose an electric battery as a place of abode.   These
branchellions always attach themselves, as it appears,
to the lower surface of the body, and not to the gills as
has been thought; and they are distinguished from all
their congeners by tufts of filaments along their sides,
which have been compared to lymphatic branchiæ.

Many naturalists have considered these curious worms
worthy of attention, and have made many interesting
observations upon them.   One of the finest memoirs
cn this subject is that of Mons. A. de Quatrefages.   We
may here mention, in connection with their mode of life,
that neither Leydig nor Quatrefages found globules of
blood in their digestive cavity.   The branchellions live
on the mucous products of the secretions of the skin, and
instead of being parasites, we may consider them as
worms paying liberally for the room which they occupy in
their host, by maintaining his skin in good condition.
They ought rather to be classed among animals which
render service to others ; that is, among mutualists.

In the fresh waters of Europe, a little leech-like
animal, beautiful both in form and colour, fixes itself
on carps, tenches, and other Cyprinidæ ; this is the
Piscicola geometra, which also lives on the *Silurus glanis*.
They are sometimes found in such great numbers that

they form around the gills a kind of living moss, which at last kills the fish.

There are different leeches which inhabit invertebrate animals. Rang mentions a little creature of this kind in Senegal, living as a parasite upon the respiratory apparatus of an anodont. Gay discovered in Chili one of the Hirudinidæ in the pulmonary sac of an Auricula, and another on the branchiæ of a crab (*Branchiobdella Chilensis*). Mons. Blanchard has noticed a malacobdella in the branchiæ of the *Venus exoleta*; and it was known in the last century that the *Mya truncata* of our coast also lodges a malacobdella which lies always under the foot of the animal. This is the hirudinean of which we have spoken above, which is allied transitionally to the trematoda.

Together with the Hirudinidæ, we find very small worms, transparent, bristling with daggers and spikes of every form, which are found everywhere in fresh water. They are known by the name of *Naïs*. They are so completely transparent that we can see the action of all their organs through the substance of the skin. They have been the subject of several remarkable works.

They live freely among the leaves of Lemna and other aquatic plants; but there is one species much more restricted in their habitat than the others; these seek assistance from the Lemneæ, and live at their expense. It is because of this kind, of which the genus *Chœtogaster* has been formed, that we mention them here. Their long bristles are veritable halberds, which they employ with astonishing skill, both in attack and defence.

Among free parasites are found many very important

articulated animals, which neither the naturalist nor the physician ought to ignore. Some of these increase with frightful rapidity on the skin which harbours them, and their name alone is sufficient to inspire disgust, if not horror : others live like leeches at the expense of different animals, but without inhabiting them. There are many of these which follow their host everywhere, and which are dreaded not without just reason.

Of this kind are gnats, fleas, lice, bugs, and a great many others, among which we ought not to forget the acaridæ, nor those singular parasites of bats, which bear no slight resemblance to spiders swimming in the midst of the fur. Volumes might be written concerning the organization and the habits of these parasites. These small creatures inspire the naturalist with no more disgust than the earth-worm of our flower-beds, or the salamanders of marshy places. Each one plays its part according to its conformation, and the most abject in appearance is' not always the least useful.

We will select among these parasites some two-winged insects, among which there are many which suck blood. Those which are generally called flies are divided into two groups, under the name of *Nemocera* and *Brachycera;* many of these live only on blood, and are more terrible than the lion and the tiger; in many countries man can defend himself against those fierce carnivora, but he is there completely powerless and without defence against these insects.

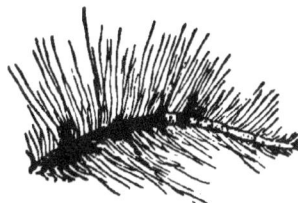

Fig. 7.—Antenna of a Gnat.

Among the Nemocera are found the gnats (*Culex*

7

pipiens), those brilliant children of the air, with fine
and slender claws, and delicate membranaceous wings,
and wearing on their heads feathery antennæ of rare
elegance.   They are known in the Old as well as in the
New World, and in southern regions it is necessary to
guard against their nightly attacks by musquito curtains.
In the Antilles they bear the name of *Maringouins*, and
in hot countries they are generally known as musquitoes.
They are also called gnats, midges, black-flies, zanzare,
&c., in different localities, but as may be supposed, these
names do not always designate the same insect.   The
musquitoes of the French colonies are often *Simulia*.
At Madagascar and the Isle of France is found the gnat
known by the name of *Bigaye*.

In Davis's Straits, in lat. 72° N., Dr. Bessels, on
board the *Polaris*, was obliged to interrupt his observations on account of these insects.   A great number of
them have been seen up to the 81st degree of latitude.
Besides gnats, there were also found *Chironomi, Corethræ*,
and *Trichoceræ*.   As Dr. Bessels was able to save from
the *Polaris* some small collections of insects, we shall
soon know the names of the species which live in these
high latitudes.   It is said that the Esquimaux and.
the Lapps cover their skin with a coating of grease, not
only to lessen the effect of the cold, but to defend themselves from the stings of gnats.

" The gnat is a plague from June till the first frosts,"
says Mons. Thoulet, speaking of his abode among the
Chippeways.   " It renders the country almost uninhabit-
able ; and one is so exhausted by this suffering, which
does not cease by night or by day, and by the loss of
blood through their bites, that we manage to get through

our daily task only by the force of habit; we can neither speak nor think. When the musquitoes disappear, the ' black-flies' come : the musquito pumps up a drop of blood and flies away; the black-fly bites and makes a wound which continues to bleed."

De Saussure has alluded to curious relations which exist in Mexico between a bird, a beast, and an insect. " Bulls bury themselves in the mud," says this learned traveller, "in order to avoid the attacks of gnats, leaving in the air only the tip of their nostrils, on which a beautiful bird, the Commander, posts himself, in this position the Commander watches for the *Maringouin* which is bold enough to enter the nostrils of the animal."

Gnats are parasites in the same manner as leeches, since, like them, they suck the blood, and live at the expense of others. There is, however, this difference, that the females only are greedy of blood; if this fail them, they live, like the males, on the juices of flowers. Another difference is that they are completely harmless till they have wings, and though they live long under their first form, in damp earth or in water, the duration of their life as perfect insects is of short duration.

We need not trouble ourselves about the active larvæ which swarm in stagnant water, nor the chrysalids which float immovable in their natural sepulchre. We give on the next page a representation of a larva of the gnat. The females alone pierce the skin by means of an auger with teeth at the end; they suck the blood, and before they fly away, distil a liquid venom into the wound. This bite seems to have an anæsthetic effect, which does not cause it to be felt till some time after.

The little spot around the wound appears as if affected by chloroform.

Fig. 8.—Gnat (*culex pipiens*) larva and nymph. (Blanchard.

These parasites repay by an unkind action the assistance which they have demanded from us.

Besides the gnats, which belong to the family of *Culicidæ*, there are also the *Ceratopogon*, and especially the *Simulium molestum*, known in North America under the name of *Black-flies*: " the tormenting black-flies of this country," as the Americans say. Certain Nemocera, known by the name of *Rhagio*, put to flight both man and animals.

They are very small; they get into the nostrils, and cause animals to become blind by introducing themselves into their eyes. In addition to these hurtful insects, we find others fatal to the life of animals, and which are a real plague in certain countries.

The numerous travellers who have explored the interior of Africa, have almost all spoken to us of a fly which attacks beasts of burden, and kills them in a few hours; this is the Tsetse (*Glossina morsitans*). More than one expedition has failed on account of this dipterous fly.   It was this which obliged Green to abandon his plan of reaching Libebe, by causing him to lose one after another all his beasts of burden and of draught.   The horse, the ox, and the dog are more especially attacked by this terrible fly between the 22nd and 28th degree of longitude, and the 18th and 24th of south latitude.   Happily it does not produce any effect upon man.

There is another fly in Mexico which is dangerous to man ; it is known by the name of *Musca hominivora*, or more correctly, *Lucilia hominivora*. Vercammer, a military surgeon of the Belgian army, relates that a soldier in Mexico had his glottis destroyed, and the sides and the roof of his mouth rendered ragged and torn, as if a cutting punch had been driven into those organs.   This soldier threw up with his spittle more than two hundred larvæ of this fly. We give below the figure of the larva and of the perfect insect. He had found this man sick in Michoacan, at a height of 1,866 metres, between Mexico and Morelia.

Fig. 9.—Lucilia hominivora.   Fig. 10.—Lucilia hominivora, larva.

My son-in-law, Dr. Vanlair, informs me that citric acid or the juice of lemons is efficacious in destroying these insects. Injections of this acid are thrown into the nasal fossæ.

At Brazil, in the province of Minas Geraes, they give the name of *Berne* to a fly which attacks man and cattle from the month of November until February. It deposits its eggs in the loins, the arms, the legs, or even the scrotum, without the victims perceiving it, and their presence is first shown by a redness, then by a sensation of itching, and a swelling with the formation of pus.

Among those insects which suck the blood, is one which is known by every one, the Breeze-fly, *Tabanus bovinus*. Happily it seldom attacks any animals except oxen and cows. We give a representation of the insect, the parts of the mouth, and one of the antehnæ.

In the same order of diptera are found ordinary flies, among which may be easily distinguished the three species which are here represented, and which differ as much by their external characters as by their mode of life.

Another fly also attacks horses and cattle, and occasionally even man, the *Asilus crabroniformis*, whose wounds sometimes draw blood. Martins, the birds of the twilight, which fly in flocks above the houses,

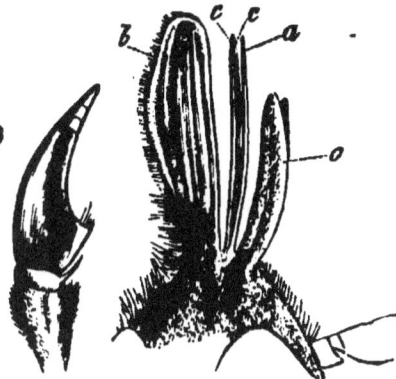

Fig. 11.—Ox-fly.   Fig. 12.—Antenna of Ox-fly.

describing circles and uttering shrill cries, are usually infested by many vermin, among which we find a fly of

Fig. 13.—Blue Fly.

considerable size, which looks much like a spider, the *Ornithomya hirundinis*. It moves about among the feathers with astonishing facility, and it is not always confined to the same bird; it quits its host to establish

Fig. 14.—Flesh Fly.   Fig. 15.—House Fly.

itself upon another, and sometimes throws itself upon man to suck his blood.

Some years ago these insects penetrated in the middle of the night through the open windows into one of the apartments of the military hospital at Louvain, and the next morning the skin of many of the patients, and especially the bed-linen, were covered with stains of blood. The physicians sent me some of these insects, not knowing whence they had come, nor whether they had been the cause of this annoyance. During the night, these Ornithomyæ had quitted their hosts to attack the soldiers.

One of these insects, the banded Syrphus (*Syrphus balteatus*), when in the larva state, seizes the rose aphides, and sucks their blood with great eagerness.

But it is not precisely a case of parasitism, when the wounds of soldiers are covered with larvæ, of which there were many sad instances in the Crimean war. There are flies which deposit their eggs in pus, as in all kinds of animal matter in a state of decomposition. It is even said that these insects, deceived by the smell of the Arum flower, will lay their eggs on the pistil. The name of *Myasis* has been given to the presence of these larvæ in a wound.

Every one knows that bats are often literally covered with vermin. Among the many parasites which attack these little animals we find, besides the acaridæ, a *Pteroptus* of great agility, which seems, as it were, to swim among the fur, and looks like a little spider or a microscopic crab. There are but few bats on which we do not find some of these, and we have sometimes seen them in such abundance, that it was impossible to touch a single hair without disturbing them. This species is usually called *Pteroptus vespertilionis*. It is constantly in motion, and glides among the fur like a mole in a sandy soil.

Together with these Pteropti lives a parasite of gigantic size, which insinuates itself among the fur with equal dexterity, and bears the name of *Nycteribia*. This has long claws like a spider, and plunges deeply into the fur. These Nycteribiæ are found only on bats. They are often associated on these animals with fleas and mites. Mr. Westwood has written a monograph upon them. Mons. Plateau, our colleague, has quite recently described a new species in the "Bulletins de l'Académie de Belgique."

Among the insects justly dreaded by man, and which follow him everywhere, is found one of the Hemiptera, known by every one under the name of bed-bug (*Cimex lectularia*). It is said that this insect was unknown in the capital of Great Britain before the fire of London in 1666. According to some entomologists, it was in-troduced into Europe in some wood that came from America. It is only necessary to make this slight refer-ence to the Cimices; their congeners are, for the most part, parasites of plants, and live on their sap.

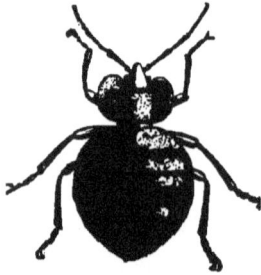

Fig. 16.—Bed-bug.

To the same order belongs the singular hemipterous insect of our ponds, the boat-fly (*Notonecta*). It has some feet suited for swimming, and others for run-ning, and it swims on its back with great rapidity. It is a dangerous neighbour for everything that has life. Always greedy of blood, it attacks great as well little animals, and sucks the blood of its victim to the last drop, so that it must be closely watched when placed in an aquarium.

Lice, concerning which we are about to add a few words, are also free parasites, and belong to a different order of insects. Their mouth is formed of a sucker contained in a sheath, without articulations; it is armed at the point with retractile hooks, within which are four bristles. They have climbing feet, terminated by pincers, with which they seize the hair of the animals on which they live; their eggs are known by the name of *nits*. We have represented in Figs. 17, 18, and 19, the complete insect, the head, the sucker, and a claw more highly magnified.

Lice are hatched at the end of five or six days, and reproduce at the end of eighteen days. Leeuwenhoek calculated that two females might become the grand-mothers of 10,000 lice in eight weeks. They are all parasites of the mammalia, and three species live at the expense of man : the louse of the head, of which Swammerdam gave a detailed description in his work entitled "Biblia Naturæ"; the body-louse, which lives on the bodies of filthy people, forms a distinct species; the third species is the louse which occasions the disease called pedicularis, or *Phthiriasis.* These insects were formerly much more common than they are at the present day. In 1825 Dr. Sichel

Fig. 17.—Louse of the Head.

published a monograph concerning them; and there appeared in the "Gazette Médicale" of 1871, a long article on the history of *Phthiriasis.*

It is stated that several great personages have fallen victims to its attack, but these observations date from a period when it was thought that they could be spontaneously originated. It is in fact difficult to believe, as it has seriously been stated, that lice have been seen to issue from the bodies of men like a spring of water from the earth. A physician of the 16th century, named Amatus Lusitanus, speaks of a great Portuguese nobleman who was so covered with lice that two of his servants were constantly occupied in collecting them and carrying them to the sea. Andrew Murray has published a memoir on the lice of the various races of men.

The name of helminthiasis has been proposed for worm disease in general, and either tæniaceous or lumbricoidian helminthiasis, according to the species which made its appearance. These parasites were considered to be formed spontaneously, and their presence constituted a pathological condition, two errors which have now been recognized, and by which the science of medicine has profited.

The *Phthirius-pubis* is another species which has been found only on white races, and attaches itself especially to the hair on the pubis.   Mons. Grimm has published in the bulletins of the Academy of St. Petersburg, an

Fig. 18.—Louse of the Head; 2, 3, sucker.

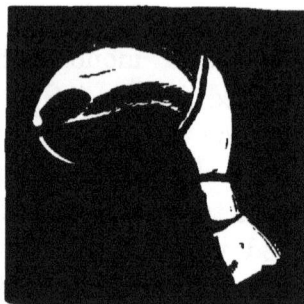

Fig. 19.—Louse of the Head, claw.

interesting memoir on the embryogeny of this insect; and, more recently, Mons. L. Landois, of Griefswald, has completely studied its habits.

We are now about to refer to certain parasitical insects whose name is usually associated with those which have preceded; they are well known by all, and attack both men and the mammalia with no less ferocity; we allude to fleas, which differ from gnats in this respect, that the male is as eager for blood as the female, and that both of them, like leeches, live by sucking it; besides, the larvæ of fleas live only on what the full-grown insects bring them, whereas the larvæ of gnats get their own living; the mother flea sucks for herself first, and then divides the spoil with her larvæ which as yet have no feet. For a long time it was thought that the fleas of different animals belonged only to a single species, and consequently that the flea of man was not different from that of a cat or a dog.

Daniel Scholten, of Amsterdam, in 1815, showed by his microscopical observations, that fleas differ from each other; and in 1832, Dugès of Montpellier, investigated the distinctive marks of the various species. The observations of Scholten may be found in "Les Materiaux pour une faune de la Néerlande," by R. T. Maitland.

The ordinary flea is called *Pulex irritans*, and especially attacks man in Europe and in North America ; it may be called a fly without wings, and, together with its congeners, it forms a distinct family under the name of *Pulicidæ*. Van Helmont treated of these insects, and gave directions for making them, just as though he were describing a recipe for pomade. At that time, naturalists supposed that certain fish could be formed spontaneously, and that nothing but fermentation was necessary in order to bring forth a crowd of living creatures from this molecular disaggregation. Fleas may, perhaps, some day find a place in the chemist's shop as well as leeches. We see no reason why homœopathic bleedings should not·be resorted to, as well as homœopathic medicines; we should certainly have more confidence in the effects of the bites of fleas, than in the efficacy of remedies subdivided into the millionth part of a grain.

Fleas differ much in size, according to the places which they inhabit. Dugès, of Montpellier, gives us a curious instance of this. He devoted himself to researches on the zoological characters of this genus, studying the four species which are the best known, the *Pulex irritans* of man, *Pulex canis* of the dog, *Pulex musculus* of the mouse, and *Pulex vespertilionis* of the bat.

Fleas of a brown colour, almost black, and of enormous size, are commonly met with on the sandy shores of the Mediterranean, at least, in the neighbourhood of Cette and Montpellier; they are more than half as large as a common fly. These are human fleas, and their presence on the sea-shore during the heats of summer is due solely to the great number of bathers of both sexes and of all classes, which lay their clothes down there. If at some future day these insects were to be placed in the rank of surgical species, it would be

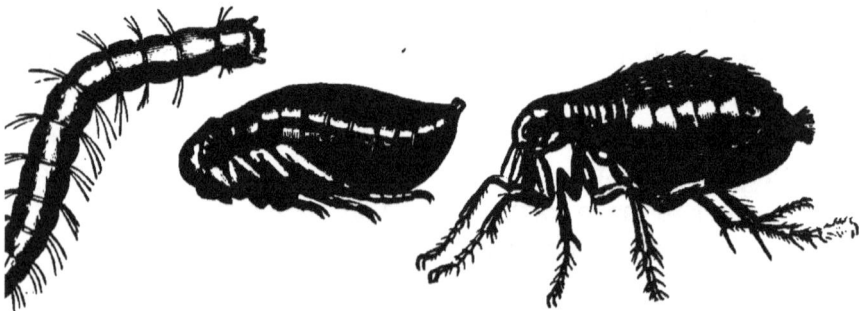

20.—Human Flea (*Pulex irritans*), after Blanchard.

necessary to resort to those shores in order to procure them; and we might suppose that, by judicious crossing, we might soon produce races that would be of real service; as yet, however, the therapeutic art has had recourse only to leeches. Since we have seen these insects harnessed and performing their exercises in public, we cannot say that the future may not reserve for us a still greater surprise.

None who saw them can have forgotten the exhibition of learned fleas made by a young lady who had sufficient patience to train them. Walckenaer saw them in Paris, and examined them with the eye of an entomologist; he relates that thirty fleas performed their feats at evening exhibitions, for admission to which the sum of sixty centimes was paid; that these fleas stood on their hind legs, armed with a pike, which was a very thin splinter of wood; some dragged a golden chariot, others a cannon with its carriage, and all were attached by a golden chain on the thighs of their hind legs.

It is curious to see how Leeuwenhoek described, two centuries ago, the history of the flea, with all its details, the accuracy of which can scarcely be surpassed. He observed their entire anatomy, as far as was possible with the instruments of his time (1694), and his descriptions are accompanied by excellent plates; he saw them copulate and lay eggs, and followed their whole development.

The finest fleas, both as to their size and form, inhabit the bats. Fleas are often found on horses. A colonel of cavalry, on his return from the frontier in 1871, sent me some of these insects, with the request that I would examine them. He added that the horses of his regiment were literally eaten up by them. It was the *Hematopinus tenuirostris*. There is a species peculiar to monkeys, which Mons. Paul Gervais has described under the generic name of *Pedicinus*.

At the commencement of the last century, a certain physician attributed the cause of almost all diseases to microscopical insects, and gave figures of ninety species which were supposed to produce, in some cases small-pox, in others rheumatism and gout, jaundice and whit-lows. Almost all these figures represent imaginary creatures. This opinion has reappeared in modern times; how many persons have been seen to smoke camphor in order to preserve themselves from the invasion of animalcules. I do not speak of the apparatus which has been contrived in order to breathe nothing but air which has been filtered and deprived of its living germs.

There are some of the articulata with four pairs of feet, a kind of microscopic spiders which require to be noticed here; these are the numerous Acari which infest many animals. Some of these wander on the surface of the skin, others in galleries under the epidermis, and many pass from one animal to another without changing their form or mode of life. There is a considerable number of them; no class of the animal kingdom is free from them, neither aquatic nor terrestrial animals, neither vertebrates nor invertebrates. These parasites belong for the most part to the same family, and cause by their presence a disease which was for a long time considered to be peculiar to the skin.

An English naturalist, Mr. George Johnson, carefully studied the parasitical and free acaridæ of Berwickshire. Mons. Ehlers has written a very interesting work, with fine illustrations, on the acaridæ of birds, published in the "Archives of Troschel." There is more than one species which lives at the expense of man, and one of

them produces a disease known in every country and at
all times under the name of
the itch; until 1830 its true
nature was still unknown.   It
is not an affection of the skin,
as was thought, but merely
the result of the presence of
these animalcules.   The director of the special Hospital
for Skin Diseases at Paris was
so fully convinced that the
acaridæ are not the cause of
the itch, that he offered a
prize to any one who could
render these insects visible.   A student of medicine, a

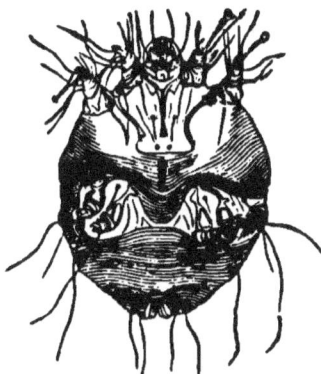

Fig. 21.—Sarcoptes scabiei, or male
acarus of the itch; the lower
surface.

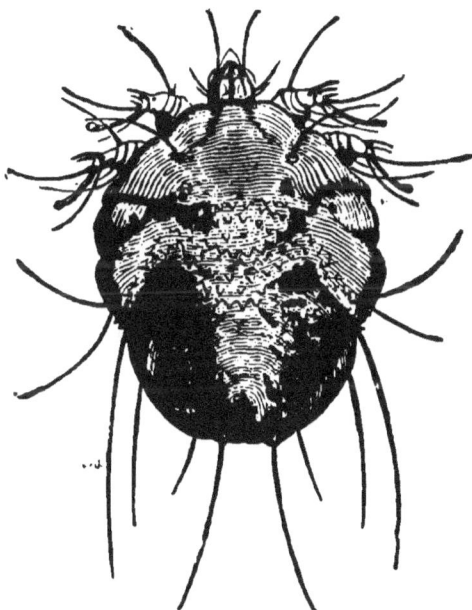

Fig. 22.—Sarcoptes scabiei, female; the upper surface.

Corsican by birth, had happened to see these itch-insects
sought for in his own country, and was the first to prove,
in 1834, the real cause of the disease. A resident
student had given, in a thesis which he sustained at
Paris before the faculty of medicine, a drawing of a
cheese-mite instead of the itch-insect, and this error
had caused it to be supposed that the species peculiar to

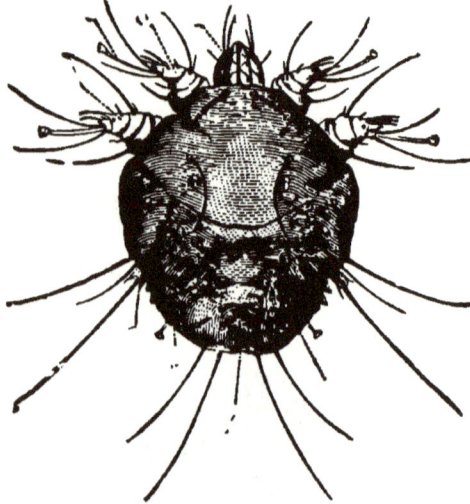

Fig. 23.—Sarcoptes scablei, male; the dorsal surface.

this disease did not exist. We give in Figures 21, 22, 23,
representations of the male and female insect, greatly
magnified.* Of course, all the treatment necessary for
the cure consists in getting rid of the animalcules and
their eggs, and in cleansing the skin and the clothes of
the patient. Petroleum oil has been judiciously prescribed in order to destroy the mite, but the remedy
which seems the most efficacious is Balsam of Peru.

* Hardy, in his *Leçons sur les maladies de la peau* (Paris 1863),
devotes a special chapter to parasitical diseases, and gives the complete
history of the itch-mite.

Most mammals have their peculiar species of acari, and the horse has two which give rise to different skin affections.   Since the presence of these animals constitutes the disorder, it may be easily caught; man may communicate it to domestic animals, and they may give it to him.   The itch-insect of man bears the name of *Sarcoptes scabici*, and no other species than those of Sarcoptes can be transferred from animals to man. These animalcules have at different times been diligently studied by many naturalists, and Dr. Füestenberg has lately published a folio volume, under the title of " Die Krätzmilben der Menschen und Thiere," with large lithographic plates, and illustrations in the text.   It is possible that the pustular disease which prevails at Sierra Leone is originated by some peculiar acarus.   Another acarus parasitical on man, the *Persian Argas*, is fortunately unknown in Europe.   It is said to be common at Miona, and prefers to attack strangers.   Its stings produce acute pain, and travellers assure us that they may be the cause of death.   This acarus remains but a short time on the person, and generally makes its appearance during the night.   It is called also the Miona bug. Fischer of Waldheim has published a very, interesting memoir on this parasite.   Justin Goudot has also observed another Argas (*A. Chinche*) which torments man in the temperate regions of Columbia.

These Arachnida, for they are articulata with four pairs of legs, often make their appearance where we should not expect to find a living organism, and naturalists, under these circumstances, have, with the best faith possible, supposed that they had seen these mites produced spontaneously without parents.   We have seen a remarkable instance of this in the *Acarus marginatus* of Hermann.   On the 18th Thermidor, an 2, they were making a *post mortem* examination at Strasburg of a man who had died of fracture of the skull, and when opening the dura mater, they saw on the corpus callosum, a mite running about which became the type of the species. The appearance of this acarus under such conditions made, as may be supposed, much noise at the time, but we should not be surprised if it had been introduced during the operation by a fly seeking to lay its eggs.

In this group is found another interesting acarus, which is developed in man in the sebaceous crypts of the nostrils.   The name of Simonea has been given to it, from Dr. Simon of Berlin, who made it his especial study. This genus leads us by its form to the *Linguatulæ*, the structure of which has been so long doubtful.   The *Simonea folliculorum* belongs to the family of the *Demodicidæ*.

The dog harbours a demodex (*D. Caninus*) which causes it to lose its hair.   Some years ago, the sheep in Belgium were attacked by one of the acaridæ, the *Ixodes reduvius*, which had been introduced from a neighbouring country, and had multiplied with frightful rapidity. Packard has given an account of an *Ixodes bovis* on the *Erethizon epixanthus*, and on the *Lepus Bairdii*, and an *Argas Americana* on cattle coming from Texas ; this was published in the sixth report of the United States' Geological survey (1873).

According to the observations of Mons. Megnin, the *Tyroglyphi*, the *Hypopi*, the *Homopi*, and the *Trichodactyli*, are transitory forms which ought not to be preserved as generic divisions among the acaridæ.   We have found on the small bat (*Pipistrella*) an acaride (*Caris elliptica*) and a new *Ixodes* which we have described in a special memoir on the parasites of the *Cheiroptera*. Mr. Lucas caught an ixodes on a dog, and kept it alive long enough distinctly to see it lay eggs which proceeded from an oviduct. These eggs formed masses attached to the abdomen of the mother.

An acarus (*Dermanyssus avium*) is found on birds, and multiplies with such rapidity that it completely exhausts those on which it has established itself. It has been seen accidentally on man. An instance is recorded of a woman who could not get rid of these parasites, because she passed every day through her henhouse in order to get to her cellar, and the frightened fowls threw down upon her a perfect shower of acaridæ. Not long ago mention was made at the Academy of Medicine at Paris, of a sarcoptes (*S. mutans*), which produces a disease among fowls, especially on the cock and hen, and which passes from these to the horse and other domestic animals. This sarcoptes prefers to live under the epidermis of the feet. Reptiles are not free from its attacks, for it is often seen on lizards and serpents. We have found a very curious one on the skin of a gecko from the south of France.

Many insects are always covered with certain species of acaridæ. Every entomologist knows that the body of the "watchman" beetle always has some of these, like little living pearls, which wander especially on the under side of the abdomen. It is the same with a small coleopterous insect that is found abundantly wherever there is any decomposing matter. Léon Dufour gave himself up to the study of some of the parasites of insects, and

mentions, among others, a species belonging to the
muscidæ, the *Limosina lugubris*, which does not measure
a line in length, and which harbours as many as fifteen
pteropti under its abdomen.

Bees, which give us their wax and their honey in
exchange for the shelter which we afford them, have a
mortal enemy, an acarus, which attaches itself to them,
not in order to gain any advantage from them, but to
cause their death. It is not so much a parasite as an
assassin, and we may be excused from describing it. We
have found acaridæ on certain polyps, the *Campanulariæ*
and *Sertulariæ* of our coasts, and some years ago we
described one which is very curious, and inhabits the
southern whale, in the midst of its Cyami and Tubi-
cinellæ. The anodonts of our ponds, as well as the

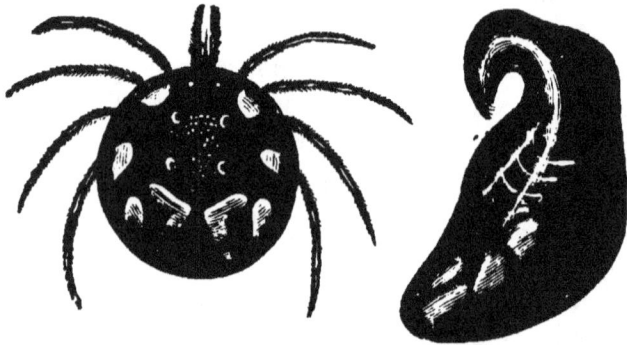

Fig. 24.—Hydrachna geographica.

Uniones usually have the skin of their feet and that of
their mantle encrusted with acari of every age, to which
the name of *Atax ypsilophora* has been given. The
species which live on the anodonts are not the same as
those which inhabit the *Uniones;* and Mons. E. Bessels,
who has so fortunately returned from his voyage to the

North Pole, on board the *Polaris*, has seen the species of the anodonts crossed with those of the *Uniones*.

There are also Arachnida which are parasitical only while young, as the *Trombidions* and certain *Hydrachnæ* (Fig. 24) which frequent aquatic animals. The *Leptus autumnalis*, known in France, at least in some localities, by the name of *Rouget*, is an acarian which throws itself upon man, and especially attaches itself to the roots of the hair : fortunately, it is only found in the country districts. The *Acarus* (*Cheyletus*) *eruditus* (Fig. 25) lives in books and collections, as well as on fruits and all kinds of bodies more or less damp, left in dark places ; it has been studied by Van Der Hoeven. Mons. Leroy de Méricourt found in pus, which

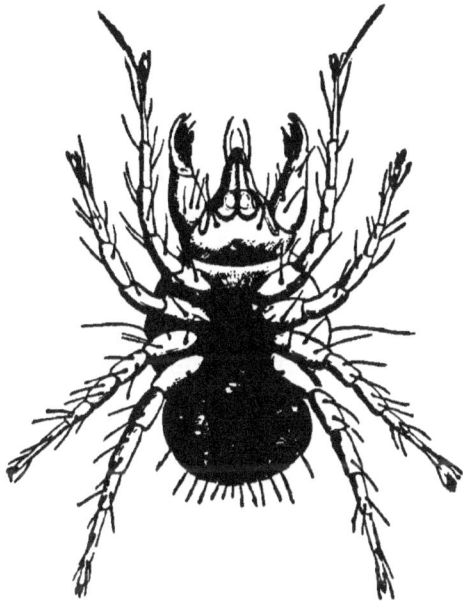

Fig. 25.—Cheyletus eruditus.

was running from the ear of a sailor, acaridæ which Mons. Robin refers to the genus *Cheyletus*, rather than to that of the *Acaropses*.

CHAPTER VII.

PARASITES FREE WHILE YOUNG.

WE have brought together in the former chapter the animals which live at the expense of their neighbours, without seeking for anything except shelter. They seize their prey as they pass, are nourished by the blood of their neighbours, but never think of establishing themselves in their organs during any period of their life. They are almost as much carnivora as parasites, and only differ from the former class because they spare the life of their victims. They are unlike ordinary parasites, since they are contented with their food alone; and their appearance from the period of their entrance into the world is that of free animals. Those whose history we are now about to sketch, live in freedom like the preceding during all the time that they are young; like them, they are completely independent during the first period of their life; but when they have arrived at mature age, when the endless cares entailed by their young ones come upon them, they change their costume and accommodate themselves as well as they can to the new lodging which they have chosen. There is often not the least resemblance between these creatures in their youth and their adult state. All these parasites have lived a joyous life

before choosing the host which is to serve them as a cell;
but though in many species we see both sexes shut
themselves up as in a cloister, some species are to be
found in which the female alone seeks for extraneous
aid; which is not surprising, since she alone undertakes
all the charge of the family, and this would be beyond
her strength, and would endanger the life of her off-
spring, if she did not receive help and protection.

The host resembles in some respects a lying-in
hospital, especially when the female alone seeks for herself a resting-place and her food, which is not always the
case. We find, in fact, in a considerable number of Lernæans, that the microscopic male passes unperceived
upon his female, and when he renounces his bachelor
life, she feeds him with her own blood. There cannot
be a more faithful husband, since he only plays the part
of a spermatophore. We find a still more curious
example in this respect, and in which the dignity of the
male is not less compromised; we refer to the Bonelliæ
which live freely in the sand, and whose males establish
themselves parasitically on the sexual organs of the
female. She herself lives by her own industry, nourishes
her husband, and alone provides for all the requirements
of maternity.

In a later part of this work, we shall mention worms
which live in freedom in damp earth, and whose direct
progeny, entirely composed of females and hermaphrodites, can only exist as parasites. These worms do not
resemble their mother but their grandmother, and if
their descent had not been traced, they would doubtless
have been taken for species entirely distinct from each
other. Thus it is not always the whole family which is

8

modified; the male often preserves all the attributes of his sex and of his youth, while the female changes entirely her appearance and her mode of motion, especially at the approach of the period when the interest of the species prevails over that of the individual.

We can nowhere find more graceful and regular forms during the whole of their early youth than those of many of these parasites; we can never see more ungraceful, we might almost say more comical, attitudes than those of the greater part of these creatures when full grown. One might take them for some misshapen excrescence, or some scrap of wasted flesh on the body of their host. A certain number of insects are found which lead this singular kind of life, but this is more especially the case among the crustaceans, particularly the copepod crustaceans. Among all these we find the most absurd recurrent forms; in fact these animals instead of carrying on their evolution, like the caterpillar which becomes a butterfly, retrograde rather than advance, and acquire an appearance and character which prevent us from recognizing their origin. Many of these are at present known, whose graceful form is so completely changed, that without referring to the study of their embryo state, one could not tell to what class they belong. Nothing remains of their organs except the sexual apparatus and a shapeless skin. These curious parasites live also on the surface of bodies, and sometimes in the cavity of the mouth; but in fishes they are most frequently found in the branchial membranes. They look like natural setons, and it is not impossible that they sometimes fulfil the same functions.

We will first examine some insects, then certain

isopode crustaceans, an order to which the Cloportidæ
(wood-lice) belong, many of which require uninterrupted
assistance; then we will turn to the Lernæans, which
surpass all the rest in their many and bizarre transformations.

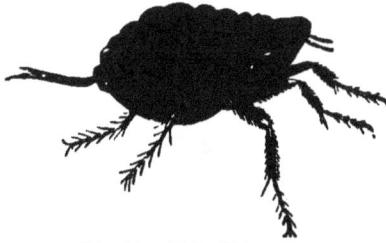

Fig. 26.—Male Chigoe.   Fig. 27.—Head of Chigoe.

We have first to speak of the Chigoe, an insect, the
female of which alone demands lodging and provisions,
the male being contented, like those of the preceding
chapter, with pillaging his victim as he passes by.   This
parasite of man inhabits South America, and has
received the name of *Pulex penetrans*, or, according to the
latest nomenclature, of *Rhyncoprion penetrans*.   It is a
very small species, which
pierces the shoes and the
clothes with its pointed beak
(Fig. 27), and penetrates
into the substance of the
skin; the male (Fig. 26) is
contented with sucking the
blood, and then resumes
its wanderings, like the
parasites of which we have
spoken in the preceding

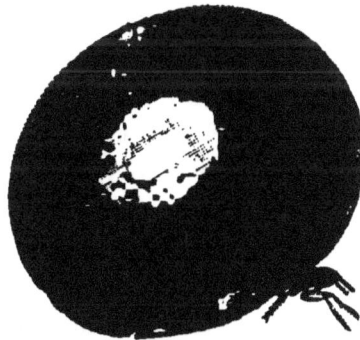

Fig. 28.—Female Chigoe.

chapter; while the female finds for herself a hiding-place, and becomes of such a monstrous size that the entire insect is nothing more than an appendage of the abdomen, as may be seen in the annexed figure. This insect is well known, since it attacks man, and usually establishes itself on his toes, but it occasionally fixes itself in the same manner on the dog, the cat, the pig, the horse, and the goat. It has also been seen upon the mule. Mons. Guyon has paid much attention to it, but we owe the last observations to Mons. Bonnet, a French navy surgeon, who passed three years in Guiana, and has ascertained that the chigoe fortunately does not extend beyond the 29th degree of south latitude. Another parasite, well known by sportsmen, is the tick. It is not an insect like the flea, but an arachnid, a kind of acarus, which passes through its last stages of development under the skin of a mammal. It is called *Ixodes ricinus*, and Professor Pachenstecher has carefully studied its organization. The ticks especially attack dogs, but are also found on the roebuck, the sheep, the hedgehog, and even on bats.

Some years ago it was propagated in an extraordinary manner on roebucks in the woods of the Duke of Arenburg, in the environs of Louvain. They are sometimes found also on man. We know of two instances: the first is that of a lady at Antwerp, who had a small tumour on her shoulder, which was removed, and enclosed a living tick. Leeuwenhoek gives an instance of a woman of the lower classes who had a tick in the middle of her stomach. Moquin-Tandon relates that Raspail found some on the head of a little girl four or five years old. He also gives an instance of a young man who, returning from hunting, found a tick under his arm; and while on the site of a

sheep market, a servant found one morning three
attached to the skin of his breast. Delegorgue speaks
of some very small reddish· ticks in Africa, which cover
the clothes by thousands, and produce distressing itching. Others are found in different parts of the globe,
and twenty-four species have been described. Several
new American Ixodes have been noticed lately by Mr.
Packard on the stag, the monax marmot, the *Lepus
palustris*, &c. These arachnida live at first in freedom in
the bushes, but after fecundation the female attacks the
first mammal which she finds in her way, and establishes
herself upon it; dogs become infested with it by running
in and out among the brushwood.

The *Argas reflexus* lives on pigeons, and is allied to
the *Ixodes*. R. Buchholz has lately studied many
new acaridæ found on different birds.

If the forms are not so varied among the isopods as
elsewhere, many among them present nevertheless the
most extraordinary appearance, the most unexpected con-
tour. Most of the parasitic isopods instal themselves
in the thoracic cavity under the carapace of a neighbour,
and make themselves contented in the small space which
remains to them. After having disposed of their
luggage, they arrange themselves scrupulously according
to the extent of the lodging which they occupy, and,
rather than interfere with the branchiæ, they raise up
the walls of the cephalothorax, thus forming a sort of
tumour which betrays the presence of the intruder.
Others are found which are not contented with a natural
cavity; they raise the scale of the skin of a fish, perforate or hollow out the true skin, or even pierce through
the walls of the abdomen, in order to establish themselves

in the intestines, still keeping up a communication with the exterior. A very common species of this class is called *Bopyrus*. We often see beautiful prawns, which are usually remarkable for their fine rose colour, exposed for sale in shop windows. If we examine them at certain seasons, especially in France, we perceive that the carapace at the side is raised ; and if we take it off with some precaution, we discover underneath an irregular flattened body, which fishermen take for a young sole on account of its shape. This is the female bopyrus. The many appendages of the thorax, the division into rings, the symmetry of the body; all have disappeared, and the claws, the traces of which are scarcely seen, are no longer similar on the right and left sides. The male remains small and independent, and preserves the livery of the order to which he belongs. On the coast of Labrador, a bopyrus behaves in the same manner towards a Mysis. We have found under the carapace of a pagurus a female bopyrus full of eggs, so much flattened that it might have been taken for a leaf accidentally introduced into this cavity.

Fritz Müller has divided the Bopyridæ in the following manner :—

1. Those which fix themselves on the appendages or in the branchial cavity of decapods; these are the Bopyri, Iones, Phryxi, Gyges, Athelgi, &c.

2. Those which live in the thoracic cavity of some Brachyuri, as the *Entoniscus.*

3. Those which live in the cirrhipeds, like the *Cryptoniscus*, as well as the *Liriopes*.

4. Those which live on copepods as true parasites, as the *Microniscus* (*M. Fuscus*).

The *Iones thoracicus*, the *Cepes distortus*, the *Gyges branchialis*, and so many others live, like the Bopyri, in the thoracic cavity of different decapod crustaceans, and the females throw off at the same time their organs of sense and all their fishing and travelling apparatus.

Rathke, a learned professor of Königsberg, was the first to notice an isopod, known under the name of *Phryxus paguri*, which lives on the stomach of a pagurus, attached to it by its back, so that the stomach of the parasite is turned, like that of the pagurus, towards the partitions of the shell. The tail with the branchial appendages is always directed towards the orifice of the shell. The male is very small and never leaves the female. The *Athelca cladophora* is another bopyrian living on the abdominal region of a pagurus, which always chooses shells infested by Alcyonia. Another bopyrian, the *Prosthetes cannelatus*, lives on the abdomen of an ordinary pagurus.

Mons. Bucholz has recently described a new kind of isopod, allied to the lyriopes, which lives on the *Hemioniscus*. This isopod fixes itself to a Balanus (*B. ovularis*), and the female preserves only four of her segments with their appendages: she had fifteen, when young. Thus she throws off nearly all her appendages which have become useless. The male of this isopod, which inhabits the bay of Christiansand, is not yet known. Another parasite of this group has been observed by Fr. Müller at Desterro, on the coast of Brazil. It bears the name of

Fig. 20.—Phryxus Rathkei. A figure of the natural size is given at the side.

Entoniscus porcellanæ. The parasite which he discovered by the side of it on the same animal, and to which he has given the name of *Lerneoniscus*, had perhaps introduced it. We have seen examples of this kind among insects. Among the rich materials which Professor Semper brought back from his voyage, there was a Porcellana, which harbours on its exterior surface a very remarkable isopod, whose recurrent development is no less decided than that of the peltogasters. Dr. Kausmann has lately described these curious organisms, to which he has given the name of *Zeuxo.* Another isopod, with a no less decided recurrent development, has received from the same naturalist the name of *Cahira Lerneodiscoïdes.*

We now come to an isopod which aims higher : he doubtless considers that cray-fish and crabs walk too slowly for him ; he therefore addresses himself to a fish, the *Puntius maculatus,* which inhabits the river Tykerang (Bandong) in Java. This isopod is called *Ichthoxenus Jellinghausii.* This isopod crustacean, living at first in the same manner as the rest, looks out for a small cyprinoid fish, thrusts itself like a trocar behind the abdominal fins, through the scaly skin, and penetrates entirely into the abdominal cavity. The male always accompanies its female. It is remarkable that she, in contradistinction to many others, preserves all the attributes of her sex. She does not change her form more than the other free crustaceans of her order, and only differs from the male in size. It is well known that in all these animals the male is always smaller than the female. Mons. Jellinghaus, who first described this crustacean, observed that all fishes which he caught had, without exception, the small ones as well as those which were larger, a couple of these parasites in their stomach. We allude to it here, but we might as well call this *Ichthoxenus* a messmate as a parasite.

On the coast of Brittany, among the many *Labri*, which are distinguished for their vivacity, and for the variety of their colours, is found a small species (*Labrus Cornubiensis*), on which is usually seen an isopod which is no less curious. It is constantly clinging to the sides of this fish, not far from the head, at the bottom of a hollow made under the scales. Naturalists have known this acolyte by Mons. Hesse's works.

This *Leposphilus* (for this is the name which has been given to it), though it does not prefer the scales to any other organ, forms a lodging for itself in the sides of this little Labrus, and takes up its abode there with its family. We cannot assert that it has chosen this refuge without any hope of returning, since both the sexes still keep their organs of locomotion.

At the last congress of German naturalists at Wiesbaden, Dr. Kossmann, who has had the opportunity of examining the rich materials brought from the Philippine Isles by Professor Semper, gave an excellent account of the result of his careful observations on some other crustaceans still more remarkable, the *Peltogasters* of which we have spoken before. In the course of this, he described an isopod with a development as completely recurrent as that of the peltogasters, whose rank among cirrhipeds is perfectly established.

Most of the inferior crustaceans require assistance from others : some might be correctly arranged as messmates, but the whole category of the Lerneans is so low in development that Cuvier placed them by the side of the helminths. These creatures possess as soon as they are born, all the attributes of their class, and wear the dress of free crustaceans; as they approach mature age, they choose a neighbour, instal themselves as conveniently as possible in one of his organs, and get rid of all their apparatus for fishing and hunting. The sexes are usually separated, and as the female is specially devoted to the cares of her progeny, she is the first to give up her liberty. Sometimes the male, not content with leaving to her all the trouble of providing for the family, demands from her his daily food, and establishes himself like a spermatophore on her sexual organs. It is only right to say that in this case, the male sex is far from being the stronger, for he is often less than the tenth or even the hundredth part of the size of the female. At last we see the female lose her claws and her swimming apparatus, while the male keeps his carapace with all his appendages of the senses and of locomotion. The difference between the two sexes is so great in some species, that it would be impossible to imagine that a brother and sister could assume such dissimilar forms, unless we had watched them from the time when they first issued from the egg. The female is a kind of puffed-out worm, and the male resembles an atrophied acarus. This explains why the female was known so long before the male, whose office is only that of reproduction. Nordmann, during his residence at Odessa, was the first to begin these researches, which have been continued by Messrs. Metzger and Claus.

It is known that the Lerneans attach themselves to their hosts by indissoluble bonds, only becoming para-

sites after they have passed their youth in complete independence, and have all possessed the graceful forms so
characteristic of the *Nauplius* and the *Zoë*. When they
first leave the egg, they swim about in freedom, but
at length some day the female, thinking of a family,
looks out for a neighbour that can give her the assistance she requires, fixes herself on his skin, and rapidly
develops till she is two or three hundred times as large

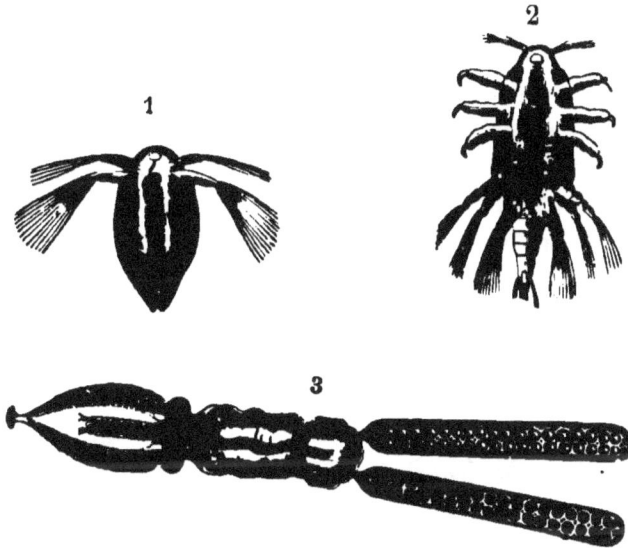

Fig. 30.—Trachellastes of the Cyprinæ. 1, larva, as it leaves the egg ; 2, larva, more
advanced ; 3, adult female, attaching itself before and behind to two ovisacs (Nordmann).

as the male; her head, her body, and her stomach
become of a monstrous size, a part of her head is often
anchylosed in the bones of her host; the lernean
remains suspended as a sort of festoon, to which are
afterwards joined two ovisacs filled with eggs. Fig. 30
is a lernean of a fresh-water fish, represented at
different periods of its existence.

The lerneans are the most remarkable of all parasites with respect to their physical degradation. They
are met with on all aquatic animals, commencing with
the cetacea, and extending to the echinodermata and
polyps; but it is especially on fishes that they are most
abundant. They live on the skin or the gills, and
sometimes establish themselves in the nostrils and on
the eye-ball. They often hang on the outside, but we
find some which hide themselves in the substance of
the skin, and have no communication with the exterior
except by a narrow orifice.

Some elegant lerneans, which resemble a living
pen, are called *Penellæ;* their head is divided into several
branches, which plunge like roots into the tissues and
even into the bones, so that the head and all the body
remain suspended, as well as the ovisac tubes, to a long
and but slightly flexible neck. They live on the body and
the eye of certain fishes; some of great size are found in
the Indian sea, but the most remarkable are those which
have been observed on the skin of some of the cetacea.

The *Penella crassicornis* lives on a hyperoodon; the
Penella balænoptera on a *Balænoptera musculus* among the
Loffoden Isles; the *Lerneoniscus nodicornis* on a dolphin ;
the great shark of the coasts of Ireland (*Scimnus
glacialis*) generally has a lernean on its eye. My son
brought from Rio de Janeiro some Scomberidæ, whose
skin is covered with penellæ ; and the charming fishes
so abundant on the Belgian coasts, which are called *Sprot*
by the fishermen of the country, often have round their
eyes strings which might be taken for marine plants, and
which are in reality only penellæ. We have found
sometimes many individuals on the same fish, stretching

from the head to the caudal region by means of their oviferous tubes, which in certain seasons acquire a pale green tint.

The true Lerneans, such as the *Lernea branchialis*, a species that was the earliest known upon the different Gadidæ, and which we have observed on the *Callionyme lyra*, greatly resemble the Penellæ, but their body and their head are much twisted, and with the coils of tubes which contain the eggs, you might take them for a ball of thread. (Fig. 31.)

The Sphyriones called *Leistera* have also a most singular form, and a new species has been recently observed on a fish from the Straits of Magellan. The *Conchoderma gracile* lives on the branchiæ of the *Maïa squinado*, the sea-spider of the Adriatic, and Mons. W. Salensky of

Fig. 31.—Lernea branchialis, attached to the gills of Morrhua luscus.

Charkow, found a copepod crustacean, the *Sphæronella Leuckarti*, in the egg-pouch of an *Amphitoë*. The latter parasite has very peculiar characters of conformation and embryonic evolution.

Among the molluscs, the Tunicates give lodging to the greater number of lerneans; in the cavity which is before the mouth, and by which the food passes, some are found which can scarcely be recognized, and which remain there to smell out a feast. The *Aplidium* of the coasts of Belgium gives lodging to some which are very curious, and which we have named *Enterocola fulgens*, on account of their colours. The *Notopterophorus* establishes itself on the body of the *Phallusia mamillaris*, and a certain number of these parasites are found on the annelids. Professor Sars of Christiania, and Claparède have carefully described them; and the latter saw on the *Spirographis Spallanzani* of the bay of Naples, a female which he called *Sabelliphilus Sarsii.* The genera *Selius, Silenium, Terebellicola, Chonephilus, Sabellacheres, Nereicola,* &c. infest all the annelids; the *Eurysilenium truncatum* lives on the *Polinoë impar,* the *Melinnacheres ergasiloïdes* on the *Melinna cristata*.

The echinodermata and the polyps are not free from lerneans; thus the *Asterochœres Lilljeborgii* fixes itself on the *Echinaster sanguinolentus,* and we have found a very beautiful species in Brittany on an Ophiurus; the *Lœmippa rubra,* allied to the *Chondracanthi,* lives upon the *Pennatula rubra,* the *Laura Girardiæ,* according to Mons. Lacaze Duthiers, feeds on an Antipathes. A *Lœmippus* (*Proteus*) lodges in the cavity of the body of the *Lobularia digitata* of Delle Chiaie; and lastly, the *Enalcyonium rubicundum* is sheltered by the *Alcyonium digitatum.*

There are certain worms which are free when young, and only become parasites at a later period of their evolution. We will give a few examples.

The Medina, or Guinea worm (*Filaria Medinensis, dracunculus*) (Fig. 32), is the terror of travellers who visit the coast of Guinea; it is common, not only on the western coast of Africa, but also in many other parts of this vast continent, and has been recently found in Turkistan and South Carolina (Mitchell). It was formerly thought that this Filaria could introduce itself directly through the skin as a microscopic embryo; but Mons. Fedschenko, after some observations made on the spot, and corroborated experimentally afterwards by Leuckart, is of opinion that this worm is transmitted by means of the Cyclops, a little fresh-water crustacean. Thus the parasite is received by means of the water which is drunk; and this remark is the more important since it will henceforth be only necessary to make use of carefully filtered water in order to guard against it. At the end of six weeks, the presence of the animal is re-vealed by tumours, the true nature of which is not ascertained at first ; then some wounds appear, caused not directly by the worm, but indirectly in consequence of the dissemination of its eggs. The Filaria at last is so entirely atro-phied that Professor Jacobson, after having seen it alive on one of

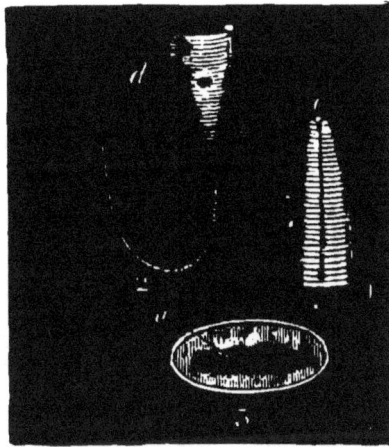

Fig. 32.—Young Filaria of Medina ; 1, Anterior extremity ; c. Mouth ; 2, Caudal extremity ; d. Anus ; 3, Section of the Body.

his patients at Copenhagen, wrote to Blainville : "This Medina worm is not really a worm, it is a sheath full of eggs." In fact, all the internal organs disappear and nothing exists there except the eggs and their embryos.

The Filaria is not allied to the *Mermis,* as was formerly thought; its organization is different, and its organs become atrophied in a very different manner. The *Gordius ornatus,* brought from the Philippines by Professor Semper, has given us an opportunity, by dif-ferent anatomical observations, to correct many errors, especially with respect to the digestive apparatus (Grenacher). The *Filaria immitis* is a species found by

Mons. Krabbe in a dog which died of a disease to which these animals are subject; it lived in the heart, and twelve individuals, ten females 'and two males, were found to be lodged there. Mons. Bap. Molin has published a monograph on the Filariæ, giving the characters of 152 species met with in molluscs, fishes, amphibians, reptiles, birds, and mammals: it seems evident that many species have been confounded under the same name.

A small worm, of the size of a slender pin, but much shorter, lives in a manner somewhat analogous to that which we have before described. It is known under the name of *Leptodera*. In order to find it, we have only to search in the woods for the first snail that we meet with, which is distinguished by its orange or black colour : if we prick with a pin the fleshy foot of the mollusc, we shall see torrents of round worms come out, wriggling like microscopic serpents. These worms also leave their retreat, if we cause the foot to contract by touching it with some acid, or if we place the snail in water. The Leptoderæ are especially remarkable for two fringes which float by the side of their tail, which characteristic suggested the name given to them by Professor Schneider. These fringes so easily fall off, that the greater part of those which have become free have none of these appendages. When placed in fresh or decaying animal matter, in water or in damp earth, these worms, agamous when in the foot of the mollusc, rapidly become sexual and perfect. Thus the snail serves them as a *crèche,* and the adult worm has no need of external help when it has grown old.

Professor Pagenstecher found at Ostend, on the *Nicothoë* of the lobster, nematodes which he arranged among the Leptoderæ. This is another instance of a parasite on a parasite.

While speaking of these worms, I will allude to a nematode which I observed under very singular circum-stances. I had a considerable number of skeletons or, I should rather say, separate bones, exposed to the sun upon a roof to whiten; among these skeletons there were several hyperoodons and other cetacea. All these bones had remained for a certain time in horse-dung in order to hasten the decomposition of the soft parts. They had been in the open air for several weeks, and were slowly bleaching; it had rained nearly every day. Towards the end of the month of August, I examined some of the vertebræ, and found them quite black on the upper part. Below, I discovered a mass of syrupy matter, slightly yellow, like pus that has recently issued from a wound. The sun was shining full upon the bones at this time; looking at them more closely, I saw this pus issuing from the holes which convey nourish-ment to the substance of the vertebræ; it seemed that the inside of the bones was in full fermentation. Ex-amining it with some attention, I perceived that the whole surface was in motion; an undulatory wriggling covered it as if a ciliated skin had been stretched above the orifices. I took a little of this matter on the point of a scalpel, and observed it with the microscope, and what was my astonishment when I saw the whole mass in motion as if under the influence of a magic wand. When I slightly compressed it afterwards between two slips of glass, there remained nothing before my eyes but nematode worms of very small size wriggling over each other : I found males by the side of their females ;

in the bodies of the latter were eggs ready to be laid, and millions of embryos of every age rolling over and struggling among the full-grown worms. Is this a species of worm new to science? Is it a worm which lives in freedom here, and parasitically elsewhere? The first female which presents itself allows us to answer this question. It is not a parasitical worm, at least under this form, because each female contains only one or two eggs. Parasites have so few chances of arriving at their destination, that two young ones would not be sufficient. They must have hundreds or thousands, and then the chances are against them. This worm is evidently a *Rhabditis*, but is it that which lives in the earth, or an allied species? Future observations will perhaps enable us soon to reply to these questions. We do not think that these creatures could have been brought with the bones from the Shetland Isles; they came rather from the horse-dung, and they multiplied beyond measure in the spongy tissue of the bones, where they found good cheer and a convenient lodging. A worm very nearly allied to this exists in abundance in the dung of the cow, to which our regretted colleague, the Abbé E. Coemans, had directed my attention, at the time when he was studying the *Pilobolus cristallinus*.

That which decided us to make mention of the nematode of the bones, is the singular history of an ascaris of the frog, whose young ones resemble their parents neither in size, form, or manner of life. There is one generation which can provide for themselves, and is composed of males and females; and another which requires assistance, and only consists of females; unless, indeed, those of the male sex are hidden among the eggs; we refer to the *Ascaris nigro-venosa*, the principal characters of which have been made known by Professor Leuckart. This Ascaris is a true parasite, which, when it arrives at its destination, where it finds lodging and food, leaves the lungs to go and inhabit another organ. There is nothing surprising that certain worms pass from the intestines to the stomach, mount thence to the œsophagus, and sometimes come out of the mouth; but here we have decided changes of abode in the same animal; that which shows, besides, that it is not a simple accident, is that the animal is of a different sex according to the apartment which it occupies; here, it is hermaphrodite, there it is male and female. The Linguatulæ, indeed, migrate from the peritoneum of the rabbit to the nasal fossæ of the dog: but the *Ascaris nigro-venosa* first lives in the lungs of the frog, then goes to inhabit the rectum of the batrachian, or damp earth. In the lungs it is very small and viviparous, and produces young ones which become stronger than their parents. The generation which live in the lungs are hermaphrodite, the others are diœcious; that is to say, the males and females have hermaphrodites for their parents. We have thus a mother, a simple female or hermaphrodite, very small, which produces, not eggs but young ones fully formed; and instead of living, like the mother, in the lungs, and breathing there with greater or less facility, they go and lodge in the rectum, and become, not like their mother, viviparous and hermaphrodite, but oviparous and of separate sexes. They produce in their turn a race of giants, and instead of following the example of their father or their mother, they all go and lodge in the lungs like their grandmother.

If the hermaphrodite *Ascaris nigro-renosa* alternately produces individuals of separate sexes, that is to say, if the monœcii produce diœcii, and the diœcii again monœcii, one cannot help comparing this phenomenon to digenetic generation. This is one of the striking discoveries made at the laboratory of Giessen, under the direction of Rud. Leuckart. Since then, Professor Schneider, the successor of Leuckart at the University of Giessen, has also studied these worms. Professor Leuckart wrote thus to me a few days after this discovery : " The *Ascaris nigro-venosa* presents this peculiar phenomenon, that, under the parasitical form, it produces fertile eggs without the presence of males. The embryos which proceed from the eggs become sexual worms at the end of twenty-four hours after they have left the body. This fact was first observed by M. Mecznikow, while he was working in my laboratory, and taking part in my researches. The experiment which produced this result was suggested and directed by myself, in order to continue my work on the development of the Nematodes."

We do not know if this is the place to speak of an animal which excited great attention some years ago, and which was thought to prove the transformation of animals into each other. It is a parasite which, under the form of a gasteropod, lives under peculiar conditions. It is known by the name of *Entoconcha*. Discovered by J. Müller in an echinoderm of the genus Synapta, its complete development has been vainly sought to be discovered since that time. It is evidently a gasteropod mollusc, allied to the Natices, and lives in the interior of the body of a Synapta, but we do not yet know all the phases of its evolution. It was at first thought that we had before us an echinoderm in the act of transformation. I wrote to J. Müller immediately after the discovery which he hastened to announce to me, to state that in my opinion, this was only a new instance of parasiticism; parasites are, however, so rare in this class of animals, and their mode of life is so exceptional, that one ought not to be surprised that this fact did not receive at first its true interpretation.

Professor Semper found at the Philippine Islands, in the *Holothuria edulis,* another species of Entoconcha which appears to attach itself to the anal vent of this echinoderm. He gave it the name of *Entoconcha Mulleri.* We have in it a new example of the relations which certain parasites bear to their hosts, and which are the same in both hemispheres.

The *Lichnophoræ* are infusoria, allied to the *Vorticellæ,* whose form they assume; these are "mimic species," or mocking forms, of the Trichodinæ. One species, the *Lichnophora Auerbachii* lives on the *Planaria tuberculata;* the other, the *L. Cohnii,* on the branchial membranes of the *Psyrmobranchus protensus.*

The associations in the inferior ranks of animals have functions which are of the highest importance; some to maintain harmony and health in all that possess life, others to sow the seeds of death throughout whole regions. There are, in fact, associations in the ranks of the infinitely small creatures, which sometimes have the effect of purifying and rendering more healthful, sometimes of destroying. It is among these beings, invisible to the naked eye, that we must seek for the cause of some epidemic diseases. We have here an example of what certain groups of animals are able to accomplish. The crustaceans everywhere perform the office of vultures to clear the waters from dead bodies, whether large or small, and they are in general sufficiently numerous to perform this police duty effectually. We may say that without their aid the waters along the coasts and at the mouth of rivers would grow speedily corrupt and unfit to support life. Thus it sometimes happens that when the number of these beings is insufficient, or the putrescible matter is in excess, we see the fish, the molluscs, and even the crustaceans, perish one after the other.

The last of the parasites of this category are known by the name of Gregarinæ. It appears that Gœde was the first to make observations upon them. Léon Dufour gave them the name which they still bear. They have a

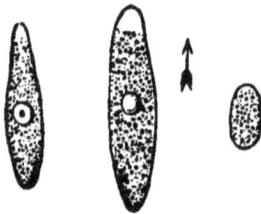

Fig. 33.—*Gregarina* of *Nemertes Gesseriensis.*

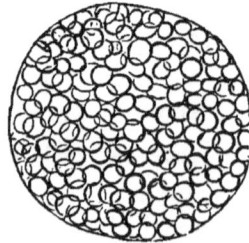

Fig. 34.—Sac with Psorospermiæ from the *Sepia officinalis.*

very simple organization, and are formed only of a cell which contains a nucleus: they live in the intestines of many invertebrate animals, especially in the articulata. Let us imagine a body, long, more or less transparent, with a smooth surface very like a spindle, which glides about in the intestines, in the midst of the liquid matter which it contains, without our being able to ascertain

the mechanism by which it moves (Fig. 33.) While
young they are encysted, and bear the name of *Psorospermiæ*. Fig. 34 represents one of these sacs of Psorospermiæ from a cephalopod.

The gregarinæ live in their perfect form chiefly in
insects, crustaceans, and worms. Fig. 35
represents a gregarina very common in
the libellulæ. The largest species inhabits
the intestines of the lobster. My son has
studied them very carefully, and published the results in the bulletins of the
Academy of Belgium.

Schneider has described a parasite
which ought, no doubt, to be placed among
the gregarinæ; it lives in the testicle, as
well as in the salivary cells, of a planaria,
the *Mesostomum Ehrenbergii;* Schneider
represents the various phases of its de-

Fig. 35.—*Stylorynchus
oligacanthus* from the
larva of the Agrion.

velopment. In the autumn of 1871, nearly
all the mesostomes perished through the presence of
these parasitical organisms : in the following year they
were rare.

Some years ago, Kölliker discovered on the spongy
bodies of molluscs, certain parasites, the nature of which
appears still as enigmatical as on the first day of their
discovery. The Würzburg professor gave them the name
of *Dicyema*. We have had for a long time in our portfolio
some observations upon them, and at the close of the
chapter "On Parasites that undergo Transformations,"
we give a representation of a Dicyema which we found in
abundance on the *Sepia officinalis* off the coast of Belgium.

CHAPTER VIII.

PARASITES THAT ARE FREE WHEN OLD.

WE are about to study in this chapter animals which seek for assistance from others while young, and are able to provide for themselves completely when they have grown old. We may compare the hosts which afford them shelter to *crèches* which receive none except new-born infants. It is generally supposed that animals known under the name of parasites are such as require assistance from their neighbours during all the stages of their existence.* This is a mistake. There are very few among them which are not able to provide for themselves during some period of their development, and they then lead an independent life. We have mentioned a certain number of them in the preceding chapter, which only seek for external assistance when they are old; we bring together, on the contrary, in this chapter, those which require help at the commencement of their life, and live at large on their own industry when they have once made their entry into the world. There are even some among

---

* The discovery of a free bothriocephalus at the bottom of a ditch caused a great sensation in the world of naturalists some years ago. It was then thought that the parasite could not exist except in the body of an animal : they could only imagine it shut up as in the cells of a gaol.

them which are richly endowed, and one would never imagine that they would have recourse to strangers in order to bring up their progeny. All their young family is usually entrusted to the care of a nurse, who lives just long enough to bring them up; she gives them convenient shelter under her roof, and often bestows upon them the last drop of her blood.

When the young one has at last abandoned her first resting-place, she begins to think seriously of Hymen ; she changes her dress and her mode of life, and seeks no more extraneous assistance till she lays her eggs. Among the animals brought up in this manner, the most remarkable are the Ichneumons, which have always attracted the notice of entomologists. These charming creatures, whose shape is delicately slender, whose transparent wings flutter with so much grace, have a less stormy youth than their boldness would induce us to suppose. As the cuckoo lays her eggs in the nest of a strange bird, the mother ichneumon deposits hers in a caterpillar full of health, by means of a long and thread-like ovipositor, so that the larvæ as soon as they are hatched, find themselves in a bath of blood and viscera, which serves them for food. The different organs palpitate under the teeth of these intruders, and the young larva grows and increases in size till it is hatched under the skin of its nurse : this skin is the cradle of the ichneumon.

The young ichneumon devours its nurse piecemeal, organ after organ ; and for fear that death should supervene too quickly, the mother takes care to chloroform the victim beforehand to make her last longer. The method which many of them adopt to get rid of their

9

young, reminds us forcibly of the turning-box in which they used formerly to place children whom they wished to be brought up by public charity ; with this difference, that young ichneumons are not only fed and taken care of by some good neighbour, but that her body itself serves them as food.

It has sometimes happened that entomologists, instead of finding beautiful butterflies produced from the caterpillars which they had reared, have had nothing hatched but a brood of ichneumons. Was it not natural then for them to dream of the transformation of species, when they saw issuing from the skin of a caterpillar, which is usually transformed into a beautiful chrysalis, a swarm of small winged flies which disperse with the rapidity of lightning ? These ichneumons discover with astonishing ingenuity the caterpillar which can bring up their young, and they often reach it with their ovipositor in the midst of a fruit, or in the substance of a branch of a tree. Every one knows the *Anobium* and other little beetles which attack wood, and live in the dark galleries which they excavate. The mother ichneumon knows perfectly how to discover the beetle which bores into our furniture, and winged ichneumons have often been seen to proceed from worm-eaten wood. It is not only caterpillars that are sought by ichneumons for the sake of their young; many kinds of larvæ of coleoptera and hemiptera, of aphides and weevils, are attacked by the mother ichneumons, which plunge their ovipositors between their articulations. These winged corsairs well know the weak points of their cuirass.

Ichneumons are therefore decidedly parasitical at this first period of their life. As they approach maturity, the

time of which varies more or less according to the
species, each ichneumon takes his departure, seeks for
booty on his own account, and passes through the last
stages of his existence at full liberty in the open air.
Nothing is more beautiful than this insect in the plenitude
of its life.' The species of the ichneumon are very
numerous.   Mons. Wesmael has devoted a part of his
life to the study of these insects.

We often ask ourselves what can be the use of these
little creatures—what good purpose can be effected by
vermin which annoy everybody?   Michelet replied to
this question when he wrote "The Insect."   "Birds,"
says the brilliant historian, "prefer to destroy those
insects which are the most injurious."   We may say the
same of those which we are now considering.   The most
common caterpillar, and that which is the most dreaded
on account of its great fecundity, is precisely that which
is more eagerly sought by the greater number of ichneumons.   No less than thirty-five kinds of these little
assassins fall on certain species, to make them serve as
a quarry to be given to their young ones.   The *Bombyx
pini* is one of the most dangerous and destructive insects
in our woods.   The ichneumons would seem to take into
consideration the too great fecundity of this moth, and
instead of one species, as is often the case, thirty-five
different species direct their attacks upon it.   It would
be indeed difficult for the mother to withdraw her young
ones from the ovipositors of so many enemies, but there
will be always enough of them remaining to keep up the
balance in this little world; the greatness of the danger
with respect to plants will be counterbalanced by the
number of ichneumons which arrest the propagation of

the caterpillars. These insects contribute more effectually to the destruction of caterpillars than all the means employed by man. To arrest the Pyralis of the vine, its cultivators encourage the little Chalcis (*Chalcis minuta*); and it has lately been recommended to introduce the acarus which attacks the *Phylloxera*, in order to lessen the number of this new pest. Do not aphides also prevent the too rapid development of certain plants? and the black species which lives on Windsor beans has doubtless suggested to the gardener that he ought to cut off the head of the plant when the flowers appear.

Some other hymenoptera may be mentioned: for example, the *Evaniadæ*, the *Chalcididæ*, as well as the *Tachinariæ*, which are remarkable for this kind of life. At the moment when the mining hymenoptera introduce into their hiding-places the insects which they have seized, and which they destine for their young ones, the Tachinariæ introduce themselves by stealth, and lay their eggs on these provisions. Each kind of tachinariæ attaches itself to a particular insect. There is one essential difference between them and ichneumons, that the females of the latter perforate the skin of their victims with a pointed instrument, and cause their eggs to penetrate to the interior of the entrails; while the mother tachinæ, less cruel, are contented to lay their eggs on the surface of the skin, and leave to the larva the care of penetrating into the interior.

In the department of the Aube, not far from Lezignan, the Tithymalis (*Euphorbia helioscopa*) grows abundantly, and the natural guest of this plant is a Sphynx. While this sphynx is still a caterpillar, a dipterous tachinaria takes possession of it to feed her young ones. For this purpose the fly establishes itself upon the back of the caterpillar, and mounted thus, without the caterpillar's suspecting the least in the world the danger that it runs, the fly inserts her larvæ to the number of ten or twelve. When she has thus deposited these, the fly goes to seek another caterpillar, like the cuckoo in search of a fresh nest every time that she lays an egg.

The young flies, left to themselves, pierce the skin of their host, and all take their place at the banquet, says Mons. Barthelemy.

After three moults the fly is completely developed, it devours the interior of the larvæ which has nourished it, pierces the skin, and the dead body of its host, which might have been its tomb, becomes, on the contrary, its cradle.

While not far off from the remains of its feast, its own skin hardens till it becomes a veritable shell, and the parasitical insect awakes, furnished with wings, ready to recommence, after a minute devoted to love, the circle in which pass the unvarying phases of its evolution.

The female of the *Scolia* attacks the larva of the large scarabæus (*Oryctes nasicornis*), which is found in tan, and pierces it with its ovipositor at the same time that it deposits an egg in the body of the gigantic larva. The larva which will proceed from the egg will suck up the fluid parts of the Oryctes while on the grass, and the skin of its victim will serve in the spring as a cradle for its transformation into a nymph.

Scolietes also attack the large oryctes which destroys the cocoa-nut trees of the Seychelles Islands. It is the same with a large species found in Madagascar.

There are around us, even in the midst of our cities, insects known under the name of Scolyti, which attracted much attention a few years ago. The trees by the side of the high roads, and even those of our boulevards, were attacked by them, and it was feared for a time that it would not be possible to arrest this new plague, which appeared simultaneously with the oidium of the vine and the parasite of the potato.

The boulevards of Brussels were planted with fine elms, and these trees were disappearing one after another. The seeds of this plague were also sown in France, in the environs of Paris. Mons. Eug. Robert had paid attention to it, and had announced to the Académie des Sciences a remedy to arrest the evil.

The regency of Brussels invited Mons. Eug. Robert to come and put in practice the means which he had recommended to destroy the scolyti; but, if I remember rightly, the death of the trees quickly followed that of the scolyti. Nature, instead of employing pitch to arrest this plague, has simpler and more expeditious means; these are, to bring forward an insect equally small, which multiplies sufficiently to keep the terrible Scolytus under. Such is the part which has devolved on the *Bracon iniator*. It simply lays its eggs in the bodies of the larvæ of the scolyti, and destroys them.

Wesmael has related a curious fact of this kind, concerning this enemy of our plantations. These little people can be well trusted to manage their own affairs. Each of these hymenoptera ascertains with an admirable instinct the place where the larvæ of the scolyti are to be found, and with its long flexible ovipositor darts an egg into the body of its victim.

It is not only caterpillars which are assailed by mortal enemies ; the eggs themselves are watched by some hymenoptera, which pierce the shell, and lay within it their own eggs. When the larvæ are hatched, the yolk and the young tissues of the legitimate proprietor serve as rations for the usurper.

In this manner, the *Ophioneuri* live, in their larva state, in the egg of the *Pieris brassica*, the cabbage butter-fly so abundant in our gardens; without this police establishment they would multiply immoderately, and our kitchen gardens would suffer still more from the ravages of these caterpillars.

It is in vain for insects to lay their eggs in the middle of fruits, or in the substance of a leaf or a branch; there will be always some hymenopterous insect which, guided by its marvellous instinct, will pierce them with its ovipositor, and reach them without their even perceiving it.

In the substance of those beautiful leaves of the water-lily which cover our ponds in summer, we often see a charming insect, known by the name of *Agrion virgo*, or damsel dragon-fly, a name given to it on account of its graceful attitudes and its elegant appear-ance. We observe this insect deposit its eggs with great prudence, fully persuaded that they are safe in the midst of the water ; but the poor neuroptera reckons without its host. An hymenopterous insect, named *Polynema*, is there, watching every movement of ·the Agrion ; and as soon as the latter has laid an egg, the Polynema darts down like a bird of prey on its victim, pierces it, and deposits its own egg in the interior. The egg of the wounded agrion will hatch a polynema. The cuckoo acts with less cruelty, since she is contented to lay her eggs by the side of those which occupy the nest.

Remarkable examples of the refinement of cruelty and of gluttony are to be found in this little animal world. It is not enough that some among them feed on the entrails of their young neighbours; there are wasps which, in order to make the agony last longer, place by the side of the eggs which they lay, chloroformed flies, which wait patiently for the time when they can yield themselves up, still palpitating, to these young tyrants. The days, the hours, perhaps even the minutes, are scrupulously reckoned for the preparation of this living morsel. As the process of hatching proceeds, the repast acquires properties more and more adapted to the age of the young wasps.

The *Sphex* is not less cruel. Some of the insects which are found in South America attack, not the young ones, but those which are grown up, and snatch spiders from their webs as slave-hunters carry off negroes from the wood; they garotte them, and cram them into narrow cells, after having chloroformed them to preserve them more effectually. These spiders, retaining enough life not to lose their nutritious qualities, become the easy prey of the larvæ of the Sphex. The mother of these hymenoptera takes care to deposit her eggs, as well as the living booty, in such a manner that the larvæ, at the moment of being hatched, live in abundance. These young larvæ, white and without feet, are dainty enough to reject any other kind of food. This is an act of cruelty which resembles that of the ichneumon, to which it may well be compared.

The *Platygasters*, another kind of hymenopterous

insects, show their cruelty in a different manner; they live in the bodies of the larvæ of *Cecidomyæ* which are lodged in the rolled leaves of the Salix, and suck the blood of their victims.

Other insects, known by the name of *Meloïdeæ*, adopt a different plan. Their larvæ havé been long known by the name of *bee-lice*; but they had not been recognized in the perfect state, as the larvæ did not resemble their parents.

These insects undergo four different moults before they become nymphs, and at each moult their appearance is completely changed. It may be easily understood that it was long before these little beings were recognized behind their masks.

This is the manner in which they ravage our flower-beds. While they still wear the dress of larvæ, they cling to certain female hymenoptera which they know very well; and being fully assured that the door would be shut in their face if they presented themselves openly, they enter, on their neighbour's back, the galleries where their housekeeping is carried on, and at the instant that the female host lays an egg in a cell of honey, the young Meloë glides in with it, and allows itself to be shut in. During this time it continues its metamorphosis, lying in a lake of honey; it devours it all at its ease, caring nothing for the provision laid up for the hymenoptera which introduced it. It is a brigand who, having secreted himself in the carriage of a rich neighbour, introduces himself on his shoulders into his children's bed-chamber, assassinates them, and grows fat on the provisions destined for his victims.

" The *Sitaris*, the *Meloë*, and apparently other Meloedeæ, if not all of them, are, when young, parasites of certain hymenoptera," says Mons. Fabri, who has watched with rare sagacity the obscure and interesting habits of these microscopic assassins.

The *Sitaris humeralis* has a progressive development at first, a recurrent one afterwards, and then again it becomes progressive.

Aphides which are not yet full grown, and which arrest the exuberant vegetation of certain plants, are in their turn attacked by an insect which is by no means lukewarm in its proceedings. A small species of cynips (*Allotria victrix*) lays its eggs, like an ichneumon, in the body of a rose aphis, and multiplies rapidly at their expense. (Westwood).

There are certain flies which are not more delicate in their mode of life than the preceding insects. We allude to the *Œstri*. We give the representation of the species which attacks the horse.

Hinder part.   36,—Œstrus of the Horse   Anterior part.

Instead of making their attacks on those of their own class, the gadflies prefer to instal themselves on mammals and sometimes even on man. Fortunately their wants are not very great; they are contented with a little. Their presence can at most only cause some uneasiness, or some trifling functional trouble.

The œstri are dipterous like ordinary flies; but instead of passing their youth on some waste organic matter, they live in the nostrils or the stomach of some hairy animal, and undergo all their metamorphoses in the interior of its body.

Thus they pass all their youth in a *crèche*; but when they have reached the adult state, they get their own living in freedom.

These œstri especially attack herbivorous mammals, and the terms *gastricola, cuticola,* and *cavicola*, sufficiently indicate the places which they inhabit; the first kind lodging in the stomach, the second frequenting the skin, and the third establishing themselves in some of the cavities of the body.

Dr. Livingstone doubtless alludes to some kinds of œstri when he mentioned the numerous intestinal worms which infest animals in Southern Africa:

"All the wild animals," says the celebrated traveller, "are subject to intestinal worms. I have observed bunches of a tape-like thread-worm and short worms of enlarged sizes in the rhinoceros. The zebras and elephants are seldom without them, and a thread-worm may often be seen under the peritoneum of these animals. Short red larvæ, which convey a stinging sensation to the hand, are seen clustering round the trachea of this animal, at the back of the throat; others are seen in the frontal sinus of antelopes; and curious flat leech-like worms are found in the stomachs of leches" (a new species of antelope).*

* Missionary Travels in South Africa, p. 136.

A species, peculiar to the horse in Europe, usually lives in its stomach in summer; and when its development is complete, the winged insect follows the course of the food, and goes out from the anus to breathe the open air. The mother fly, excited by the sentiment of maternity, flies round the breast of the first horse that she meets, and lays her eggs there on some hairs which are not beyond reach of the animal's tongue. The horse wishing to get rid of these foreign bodies, licks them off, and thus they are introduced into the mouth, and from the tongue pass to the stomach. These eggs are hatched in the midst of the gastric juice, the larvæ leave them, and the young gadflies find in the juices of the stomach the milk which serves to nourish them. These larvæ pass through their metamorphoses in the stomach, and when the young fly has assumed its perfect form, with its delicate wings, its sucker, and its facetted eyes, it leaves the stomach, follows the path traced by the food, arrives some fine day at the rectum, presents itself at the place of exit, and takes its flight. Thus the fly can take its journey through the intestines on a portion of the digested food.

When she has once taken her flight she is very near the end of her life, and after a moment of love she gives up her place to others.

There is another gadfly which finds a *crèche* in the sheep; but instead of lodging in its stomach, it instals itself in the nostrils, which are more easily reached. This second species goes through its evolutions in the vestibule.

This is the species which sometimes introduces itself into the body of man. Many instances of this have been known, and our late colleague Spring gave a very interesting account of one of them in the bulletins of the Belgian Academy.

A gadfly found at Cayenne is distinguished by the name of the Macaco Worm; it belongs to the genus *Cuterebra*, and usually attacks the skin of oxen and dogs in South America. It is accidentally found sometimes on man. This is the *Cuterebra noxialis*. We here give the representation of it.

There is also a gadfly on the ox.

Professor Joly has devoted himself to zoological researches on Œstridæ in general. Professor Schroeder Vander Kolken, in Holland, and Mons. Brauer, in Austria, have studied them with great success.

The *Hippoboscus* is a fly which is very greedy of blood, and attaches itself to horses and oxen, especially under the tail, in the parts where there is less hair. It sometimes also attacks man.

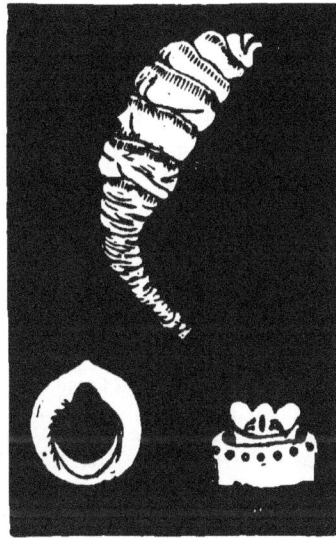

Fig. 37.—Macaco Worm.

The Hippoboscus lives on the horse, and an allied species, of which a different genus has been formed, lives on bats (*Strebla vespertilionis*) in South America. Mons. Von Baër noticed hippobosci on the elan, during his residence in Königsberg.

Many other insects live and develop themselves at the expense of their nearest neighbours.

Travellers since Azara's time assure us that Uruguay contains but few oxen and horses, because a fly exists in that country which lays its eggs in the navel of these animals at the moment of their birth. These animals, on the contrary, are abundant in Paraguay. In order to increase their number in Uruguay, it would be necessary to favour the multiplication of birds or insects which make war on these flies, either in the larval or the sexual state.

Diptera, known by the name of *Conops*, pass their first three changes in the soft parts of drone-bees. Dumeril had formerly suspected, from the curvature of the abdomen, that the Conops lays its eggs in the body of some other insect. Lachat and Victor Audouin have given an instance of this in the "Journal de Physique."

Thus the Conops, in its larval state, inhabits the abdomen of drones or other hymenoptera; the *Echinomyæ* are developed within various lepidoptera when in the state of caterpillars or chrysalids; there are even some which live on flesh, and prefer that which is in a state of incipient putrefaction.

We may also speak, in this category, of animals which seek assistance, while young, from neighbours of whom they take advantage during their life, and utilize them even after their death; these are insects of various orders. They are in general more cruel than beasts of prey, which often contend on equal terms with their victims. Here we have an enemy which furtively introduces itself into its neighbour, who is nearly sucked dry before he suspects the danger to which he is exposed. He harbours unawares the assassin who is about to murder him. This is the refinement of cruelty.

The *Melophagus* of the sheep is a wingless dipterous insect, like the *Lipoptena* of the stag. We give figures of these two curious insects.

Fig. 38.—Melophagus ovis.   Fig. 39.—Lipoptena of the stag.

The *Stratiome chameleon* pays visits to flowers to seek for insects, on whose blood it feeds. Its very elongated larva lives in stagnant water.

We have now to mention in the following passages parasites much less cruel in general, and which receive with greater delicacy the hospitality which is afforded them. We refer to some worms which pass, not their youth, but their mature age in the body of a neighbour, and use their host not as a *crèche*, but as a lying-in hospital.

Their early youth is passed in freedom, but they soon give birth to a numerous progeny. The fate of the male is unknown; as to the female, she introduces herself in a microscopic state into the body of a neighbour, is developed there till she arrives at sexual maturity, and then quits her retreat to go and scatter her eggs..

It appears, however, that these females are obliged to seek assistance from insects; but before they enter this living asylum, the male, which is not yet known, ensures by his fecundation the preservation of the species.

We often find in summer in puddles of water, thin worms, which are sometimes a foot long, resembling a violin string, and have for a long time puzzled naturalists. They are known by the name of *Gordius*, and have lately been very carefully studied, both with reference to their organization, to their mode of life, and their development. We give here the figure of a Gordius of the natural size. The *Mermis*, like the Gordius, passes its youth in the body of certain insects, and leaves its living cradle to scatter its eggs abroad. In this case, the embryos themselves go to seek for their host, and unlike the ichneumons, they use

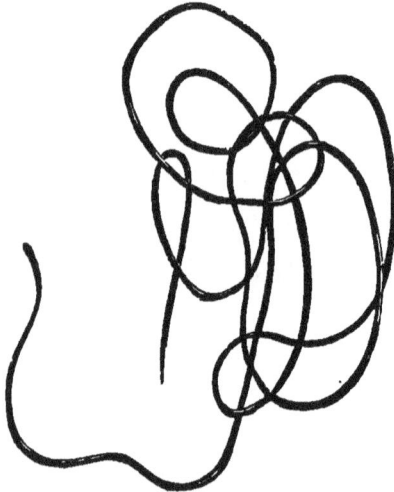

Fig. 40.—Gordius aquaticus, natural size.

them with moderation. The life of the host is never compromised, and no functional disturbance is observed, notwithstanding the enormous size of the worm.

The Mermis is especially found after a heavy shower; some kinds of *Filaria* are also more common when it rains. Under the title of "Notes on the Appearance of Worms after a Shower of Rain," I communicated to the Academy of Belgium some observations on these creatures, and these observations were recorded in the bulletins.

Some years ago they brought me one morning, after a shower of rain, a quantity of worms, four or five inches in length, very thin, and twisted round each other, which had been collected in the morning, on the flower borders of several gardens within the city. It was thought that there had been a shower of worms in the night.

There was not one male worm among three hundred; all were full of eggs, and the young ones were already wriggling about within them.

Whence come they? said I, in my article. Have they fallen from the sky completely formed? It is evident that they have not been developed on the ground where they have been found; it is not less evident that they appeared suddenly on the borders. Did they come from within the bodies of certain insects which they have quitted, on account of the rain which had fallen? These worms, in fact, had completed their parasitical stage in the bodies of their hosts, and the great drought which had continued for many weeks prevented their resuming their first course of existence. It was the sudden emancipation of so many worms at once which had attracted the attention of gardeners: earwigs, cockchafers, and many other insects give them shelter during the time of this strange gestation.

It is known, by the observations of Siebold, that the eggs of the Mermis, laid during the winter, produce in the following spring embryos which live in damp earth. They immediately seek the larvæ of insects, perforate their skin, and develop themselves there without becoming encysted. After this, they again pass through the skin of their host, return to the damp earth, where they change their skin, are fecundated, and lay eggs. The larvæ of *Mermis albicans* especially resort to caterpillars, or the larvæ of the coleoptera, orthoptera, or diptera, and even to a mollusc, the *Succinea amphibia.*

Professor Meissner, and more especially Dr. Grenacher, professor at Göttingen, have made known to us the structure of the Gordius. The *Gordius bifurcus* produces embryos at the end of a month; these embryos perforate their shell by means of their beak, become free in the damp earth, and introduce themselves through the skin into the perigastric cavity of certain larvæ. The sexual worm again becomes free. If we may believe Mons. Villot, who has made recent observations on the Mermis and the Gordius, the latter alone pass through complete metamorphoses; they assume three different forms, and change their habitation three times. Their first abode must be in the water, or in the larva of a dipterous insect, as a free embryo; the second in the larval state, in the intestines of a fish; and the third, like the first, in a sexual state.

To judge by some specimens of gordius brought from India, these curious parasites exist not in Europe only; they have been found in different parts of the world, and they lead everywhere the same kind of life.

They have been found in Calcutta in the *Hapale;* in the Philippine Islands in a *Mantis,* and the museum of Hamburg possesses some from Venezuela, which came from the body of a *Blatta.*

These worms, when they approach the adult and sexual age, lose their various external organs, and are so completely modified with respect to their organization, that at last they are merely a case for eggs. They are so entirely egg-cases, in which the digestive tube and the other organs disappear in proportion as the sexual organs are developed, that many naturalists have taken these worms for a simple ovisac. This has also been the case with the *Nematobothrium* of the fish known under the name of the eagle-fish; it has been taken by an eminent naturalist for a nest of psorospermiæ.

There are also worms which take refuge in plants, and live at their expense, as if they were in an insect. One of the most remarkable is that which attacks corn, and produces the disease known by the name of smut, the corn eel (*Anguillulina tritici.*) It is a very small and thin cylindrical worm, which dries up completely with the grain of corn which has nourished it, and which can remain for an indefinite period without dying, in a state resembling dust. Every time that it is moistened, it resumes its activity. This return to life has been compared to a kind of resurrection.

Mons. Davaine has studied this worm with great care; he has made known the different phases of its development, and the manner in which it introduces itself into the plant and the grain. Needham, in his "New Discoveries made with the Microscope," (1747) gives a whole chapter to these microscopic eels.

The larvæ of the *Anguillula scandens* are dried in the galls inhabited by the mother. As soon as these galls fall and grow moist, the larvæ revive, and abandon their cradle to live in freedom. Soon after this, they go in search of their plant, take it by storm, and penetrate into the tissues before the period of fecundation; having become sexual in the interval, these microscopic nematodes lay their eggs in a nest formed at the expense of the plant.

Another species lives in the *dipsacus*, in which also

it produces disease (*Anguillulina dipsaci*). It attacks
the flowers, and remains on them without signs of life
till the moment that they are moistened. The vinegar
eel is another nematode worm which has some affinity
with the preceding ones. It has been considered a
Rachitis.

There exists also a river species; but have not
different worms been confounded under this name?
Many species live in brackish water, and these are
remarkable for the presence of bristles on their heads,
and by very distinct eyes.

CHAPTER IX.

PARASITES THAT UNDERGO TRANSMIGRATIONS AND METAMORPHOSES.

A CERTAIN number of parasites establish themselves at first in an animal which serves as a *crèche*, then in a second which serves as a lying-in hospital. This passage from one animal to another is described under the name of transmigration. In general, the entire *crèche* with its nurslings passes into the lying-in asylum. The *crèche* is always represented by an animal which feeds on vegetable diet, which is destined for one which is carnivorous: the lying-in asylum is represented by the latter. The mouse is the *crèche* which will pass with all its clients into the cat which eats it.

If we were treating of plants, we should say that in the first host they are developed, and in the second they blossom. The plant, like the animal, is agamous as long as the flower and the sexual organs have not made their appearance.

The animal which migrates usually undergoes a complete change in passing from one abode to another; it is agamous in the first instance, that is to say, without sex, swathed and covered with a padded cap like a nursling; in its last stage it is, on the contrary, endued with all its sexual attributes.

In the *crèche* the parasite is on its passage from one station to another, and that which arrives at the lying-in asylum has reached the end of its journey and is at home. We have proposed to give it the name of *Nostosite*, as distinguished from that which only inhabits its host for a time. We may also remark that the same animal may give lodging to these two kinds of parasites. It is thus that the rabbit harbours in its peritoneum passengers which are only at home in the dog; and, independently of these passengers (these strangers may we say?), it lodges in its intestines a sexual tænoid worm. The first is a *Xenosite*, the second a *Nostosite*. The mouse, in the same manner, gives lodging to passengers under the name of *Cysticerci*, which are destined to the cat in order to become *Tæniæ.*

We might call the rabbit or the mouse which harbours worms *in transitu*, the stage coach; more especially as from time to time there are some which miss it, and are consequently lost in their peregrinations.

This stage-coach is the intermediate host, the *Zwischenwirth* of German helminthologists, which is always an animal with a vegetable diet; the final host is generally a carnivore : it is by means of the vegetable feeder, the grazing or herbivorous animal, that the stranger parasite introduces itself.

The result of this is, that the carnivore receives into its house, every time that it devours its prey, all the parasitical inmates of the latter, and the walls of its digestive canal form the soil in which are implanted all the worms which can take root there. The tissues of the prey are triturated and digested, but the worms which it encloses escape the action of the gastric juice, and are set at liberty in the stomach. The stomach of of the carnivorous animal is a sieve through which thousands of parasites are often introduced at each repast, and fishes lodge many which often pass from one stomach to another. Their whole life is spent in these migrations; they are travellers who have their abode in railway carriages, and never take their departure at the stations.

Each stomach is, in fact, a station, very frequently quite filled with merchandise, which disappears with the station itself by the next train. Happy are those who find themselves in a carriage safely on the rails towards its destination. Many are called but few chosen. How many journeys some of these travellers have to take before they find their host!

It is often very interesting to open a fish which has made a good meal; its stomach and intestines contain, first of all, the usual worms; the half-digested prey, in its turn, encloses some; and it is not rare to find besides them the parasites of those which were swallowed together with their host.

The animal is usually attacked in its youth by the parasites which it harbours all its life. In order to know the inhabitants of some fishes, we must examine them shortly after they are hatched.

In the *crèche* the parasite occupies an organ which is closed, and without communication with the outer world; it inhabits the garret of its first host; in its last host, which represents the maternity asylum, it dwells, on the contrary, in the largest apartments, and never ceases to be in direct communication with the exterior. Thus, in the first animal, it is often completely immovable and under a form which we have named *scolex;* in the latter it moves freely, and has, in addition to sexual organs, those which are proper to this condition which we have called *Proglottis*. Thus these parasites undergo metamorphoses.

For a long time, metamorphoses seemed to be the attributes of frogs and insects exclusively. In the class of worms, in which they are complicated with the change of hosts, they much surpass in reality the most brilliant and extravagant fictions of the poets. The phenomena of these transmigrations were completely unknown before our researches were made. If some naturalists, like Abildgaard or Pallas, suspected their existence, it was rather by accident, and the experiments to which they devoted themselves were all unfavourable to their suppositions.

The knowledge of these transmigrations has at the same time dispersed the latest illusions of the partisans of spontaneous generation ; it was the more difficult to explain the presence of worms in enclosed organs, since these worms were always without sex. By the same means, we have ascertained the true prophylactic treatment, and thus discountenanced the numerous anthelminthic remedies which had often caused more serious accidents than the parasites themselves.

When it was considered that parasites were the result of an especial degeneration of some of the intestinal papillæ, the physician would at once consider that there was some morbid condition, and we can understand that all his efforts would be employed against the enemy which had arisen. Now it is known that every healthy animal living in freedom contains parasites almost as invariably as the organs which support its life; and it is not a matter of doubt to us that parasites often play their allotted part in the economy; their absence as well as their presence may be the cause of inconvenience. We should not even be astonished if the administration of certain worms internally should be prescribed as a remedy. Have we not known the time when all maladies were supposed to yield to the action of leeches, and do we not see the good effects of their application? There are many kinds of parasites, and their therapeutic effect may, perhaps, in future, form an interesting subject of study.

To speak at the present time of a verminous temperament would be scientific heresy, an anachronism; this shows the progress that we have made of late years. Valenciennes was permitted to employ this language at the Academy of Sciences in Paris not twenty years ago, and Lamarck wrote thus in his standard work on invertebrate animals, in the beginning of this century: "It is very certain that there exist in a great many animals, and even in man, intestinal worms; some of which are formed there, others are born and all live there, multiplying more or less, without any of these worms showing themselves externally, or being able to live elsewhere.

"During so many centuries that observations have been made, well-ascertained species of intestinal worms have been found nowhere else than in the bodies of animals. We are now authorized to believe that there are *innate* worms, or such as are produced by spontaneous generation, and that these are modified from time to time; this is at present the opinion of the most enlightened observers."

10

Thus it was considered by Lamarck that parasitical worms are only found in the bodies of animals, and are actually produced there.

Can it be believed that such ideas were put forward by zoologists of the highest merit? and ought we to feel surprised that the theory of spontaneous generation was so long taught in the physiological schools?

A book published in 1859 was entitled, "Hetero-genesis, or a Treatise on Spontaneous Generation." The author gives the clue to the origin of his errors in the second line of his preface, in which he says : "When, *by meditation*, it was evident to me that spontaneous generation was one of the means employed by matter for the reproduction of living beings." . . . . According to this philosopher, science is, therefore, not the general-ization of facts, but these facts must serve to prop up the theories or hypotheses invented in the silence of the study. This passage of his work shows us that he was no more able to yield to the evidence of experiments made on worms, than to those of Pasteur on the infusoria.

It may be related to the honour of the illustrious Baer, that, from the year 1817, during his stay at Königsberg, he took up arms against this hypothesis, and never ceased to combat it, till evidence succeeded in opening the eyes of the most obstinate.

The worms which present the most remarkable phenomena of transformations, accompanied by metamor-phoses, are the Distomians and Cestodes, flat worms, which we will consider in the first place.

Trematode worms include a certain number of large and beautiful parasites which scarcely undergo any change, and are found only on the skin and the gills of certain fishes; these are the monogenetic trematodes, comprising the *Tristomidæ* and all the worms of that group, which also stand higher in their organization: we shall speak of them hereafter. The other trematodes, which are called digenetic, live on the most dissimilar animals, under the most varied forms, and, like the greater part of the cestodes, introduce themselves into the individual who is to give them shelter, only by the assistance of a host, acting as a stage-coach which serves them as a vehicle.

The principal family is that of the Distomidæ, a family *par excellence* cosmopolitan; as inconstant in their progress as capricious in the choice of their companions. Each distome resembles a small leech which has a sucker in the centre of the belly, and as this sucker was once considered to be perforated, the name of Distoma was given to them.

These parasites are the more interesting to us, from the fact that, though we are not the final resting-place of certain species, we nevertheless find them pass through us on their way. There are two species which occasionally lodge in the liver of man without being peculiar to him, for they properly belong to the sheep. Two other distomes have lately been described by Dr. Bilharz, which are fortunately only known at present in Cairo, and which are interesting, both with respect to their organization and to their manner of life.

The genealogy of the distomidæ is now generally well known; that which remains to be discovered is the itinerary of each particular species; and in several zoological laboratories experiments are daily made with certain species and the hosts which they are supposed to seek. These investigations have already yielded the best results in the laboratories of Giessen and of Leipzic, under the direction of Leuckart.

The genealogy of the distomidæ is as follows : the young distome, when it leaves the egg, is wrapped in a ciliated tunic, and, under the guise of a microscopic infusorial, it abandons itself to all the vagaries of a free and vagabond life; this is the bright period of its life. "It is a youth starting, with all the steam up, without help and without guidance, in the midst of the ocean ; if it meets an island on its passage, that is to say, the body of an aquatic larva or a mollusc, it disembarks, brings forth its young, and disappears; its purpose is fulfilled. If it find no island or continent it sinks and perishes, for it carries no provisions with it; it has no organ which permits it to take nourishment on its passage." If life is short, even in the case of a young distome, it is passed in the midst of the water : if fortune is favourable to it, it will at last meet with a living abode, where it will find all that is necessary to the comfort of a parasite.

Abundance always reigns in these living oases; and as these new colonists are really exiles, who will never again see their native country, ciliary oars are useless to them, and their descendants differ entirely from their common mother.

Under the ciliated tunic of the mother appears a daughter under the form of a bag, who is born almost at the same time as herself, and concerning whom we may quote here the words of Réaumur : " Singular and mysterious duality in unity ; two beings, living one within the other, which are still only a single individual. Has nature accustomed us to such profusion ? Do we ever see her retrograde thus from a more complicated organization to one more simple ?" That which this great observer did not dare to believe has yet been realized, and in many cases development is clearly recurrent.

Led by a marvellous instinct, and obeying an irre-vocable mission, the distomidæ, as well as the monostomidæ, and others besides them, when they claim an asylum from molluscs, introduce into the living body of their new host, not an isolated embryo, but a young animal already impregnated with a rich posterity; if she remain mistress of the situation, this posterity will forcibly invade the various organs, without any consideration whether their host may not give way under the weight of this sudden invasion.

Fig. 41 represents one of these worms which proceeds from a cili-ated embryo, and encloses by the side of its digestive tube cercariæ in different degrees of develop-ment. In front, we see one pro-vided with eyes and a tail; behind, we see others which are younger; among these ciliated embryos, wandering without guidance and

Fig. 41.—Monostomum verru-cosum, Sporocyst with Cer-cariæ. In front is the mouth, in the middle the digestive canal, and around the diges-tive canal are young ones, under the form of Cercariæ in process of development.

without a compass in the midst of their ocean, but few will reach the land, or, in other words, will find the

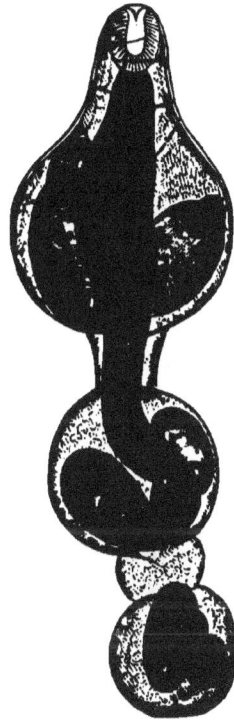

port where their progeny may prosper. This first embryonic state is that in which there are the greatest perils. When stripped of their swimming tunic, these young distomes have the form of a bag, which for a long time was called a *sporocyst*. From these sporocysts we see hundreds and thousands of young ones proceed, resembling in no respect the mother which has brought them into the world. These, in their turn, will resume a free and independent life. They are colonists whom the distome has left on a foreign land. This simple multiplication is often not sufficient for the preservation of the species ; the first sporocyst produces other similar sporocysts, and these bring into the world a rich progeny of tadpoles, which after a certain metamorphosis will become sexual distomes. These tadpoles are often well armed, and devour occasionally even the last scrap of flesh belonging to their host. They have long been known under the name of Cercariæ, which was given to them at a time when their genealogy was unknown. They are not very unlike the tadpoles of the frog (Fig. 45). The mother was only a bag with ciliæ, and sometimes with eyes. The tadpole has a distinct body, with a movable deciduous tail ; and after this falls off they have sexual organs.

The cercariæ often abandon their first host in which they have been developed, and live at liberty in the water while waiting for their final host. They are taken sometimes in the open sea. In 1849, J. Müller wrote to me from Marseilles that he had just discovered cercariæ and distomes living at liberty in the Mediterranean. Since then this illustrious naturalist has observed them again at Trieste, while pursuing his studies on the

Echinodermata, and has had the kindness to send me his original drawings of these singular parasites.

We have found both at Marseilles and at Trieste, says J. Müller, a new cercaria with a pinnate tail, and two black ocular points; its body is from one-tenth to one-sixth of a line in length, not including the tail, which is twice or two-and-a-half times as long. There is a protuberance just in front of the middle of the body. At each side of the tail there are from twelve to twenty pencils of soft bristles placed on little prominences in a transverse series of six tufts, not regularly opposed to each other. In one specimen, the tail, from its point of insertion to the posterior quarter, is provided with these bundles of bristles; and in another they are wanting entirely in the anterior half, but exist, on the contrary, on the hinder half. In a third, the bristles have partially disappeared, and are reduced to six bundles at the extremity of the tail. This tail presents traces, more or less distinct, of transverse rings. J. Müller has often seen that the distome, which proceeds from this cercaria, swims freely in the sea, and after having got rid of its tail, could be easily recognized by the two black marks which were then more diffused.

This cercaria described by J. Müller recalls to us that which was noticed by Nitzsch on fresh-water shells (*Cercaria major*) with an annulate and pinnated tail.

Claparède also took at Saint-Vaast, cercariæ the host of which he did not know. This naturalist supposed that this worm could migrate at will. He found there the same cercaria (*C. Haimeana*) on Sarsiæ and Oceaniæ, but always sexless.

The *Cercaria setifera* of J. Müller has been found

free and attached to the lower surface of some medusæ.
It exists occasionally in considerable numbers on the
internal surface of some Acalephæ of the ocean and of
the Mediterranean. Claparède has also observed another
free cercaria which bears the name of *Pachycerca*.

Some of the cercariæ are very tenacious of life; we
have kept some alive in fresh water during a whole week
in the month of November, and on the last day they
were still active (*Cercaria armata*). We sometimes find
the cercarian age passed over, and the young distomes
appear abundantly without tails in the sporocyst. We
have seen an example of this in the *Buccinum undatum*
of our coasts. This latter generation assumes in every
case a very different form from that which preceded it.

Lodged and nourished without expense in the succulent parenchyma of their victim, the cercariæ grow
rapidly, and as soon as their caudal oar is developed,
they tear asunder the membrane which encloses them,
and abandon their host in order to live freely as tadpoles. Some fine day, tired of their nomadic life, they
choose another host, get rid of their tail, fold themselves
up in a winding-sheet, like a chrysalis about to become
a butterfly, and concealed in a sac, which is designated
by the name of cyst, they wait patiently for days,
weeks, or years till their host is swallowed by the
creature intended to lodge them. The cyst is set free in
the stomach of the latter host, its envelopes are dis-
solved in the juice secreted by its enclosing membrane,
and with its whole establishment the worm recovers its
liberty in this new abode.

The encysted cercariæ pass thus with arms and baggage into the stomach of a new host. Their envelopes,

not to say their swaddling-clothes, are torn to pieces by
the gastric juice, and at the end of their stage they go
and lodge in larger apartments, more appropriate to
their new wants. The time of their celibacy is passed,
and a numerous progeny, under the form of eggs, is
prepared. In this condition they fulfil their last
mission; and if their mother, the sporocyst, knew only
the joys of agamous maternity, the cercaria which has
just become a distome appreciates all the sweetness of
sexual maternity.

The distome thus reaches the termination of its voyage and of its evolutions; it lays its eggs in the midst
of the feces of its host, and millions of animalculæ
watch for the new brood, while others wait for the visit
of the ciliated generations. The daughter distome thus
differs completely from her mother sporocyst, but she
resembles her grandmother who has lived in the same
manner as herself. Thus we have animals free and
vagabond when they leave the egg, and which swim
vigorously like infusoria without depending on others.
But the end of their life approaches, they strip themselves of their ciliated mantle, and being again closely
swathed up before they die, they seek the hospitality of
a mollusc and give birth to their numerous progeny.

We have therefore animals whose little ones in
swaddling clothes live at first at liberty, and seek for
assistance when the moment for thinking of a family
approaches. The descendants lead, like their parents, a
wandering life; and as their mother threw off her ciliated
cloak, so they abandon their oar-like tail, to think in
their turn of family cares.

To sum up all, there are in the life circle of a distomian two distinct forms, which begin and end in the same manner, the first putting forth a progeny by means of buds, the second by eggs. There is alternation of form, on account of the double multiplication (digenesis) and migration through several individuals. In other words, the young distome, before it reaches its destination, must change its train many times, and it wears in each carriage a different costume. We can easily understand how difficult it is to recognize this travelling distomian, as it changes continually its railway-train and its dress, and what sagacity must have been employed by naturalists in order not to lose its track.

We may give more than one description of the distomian embryo as it leaves its sporocyst. Is it a mother and an enclosed daughter, as is the case with aphides, or is the ciliated envelope merely a cloak? We think that the latter is the true interpretation. The ciliated mantle which the embryo loses, is a skin which has been thrown off in moulting, a simple effect of age.

Thus we find in the complete evolution of a distome an organic and a sexual age, a true alternation; the agamous age undergoes a true moulting, the sexual age a metamorphosis.

We have before considered the embryo as mother and daughter coming into the world together, as we see among the aphides; or the mother, daughter, and granddaughter are born together like twins; so that if the mother or the daughter meet with an accident during parturition, the granddaughter may be born before her mother, and even before her grandmother.

We are now about to study some of these mysterious travellers which have given so much trouble to naturalists to discover their abode and determine their identity. Considering the number of observers who have mentioned these distomes, it is evident that these parasites must be very common. We find the names of Ruysch, Leeuwenhoek, Swammerdam, Camper, Houttuyn, Mulder, Heide, Biddloo, Snellen, etc., among the naturalists who have made them a subject of study. In our own day, the writers who have explored this territory are so numerous that we should require more than a page simply to give their names.

Distomes frequent, with few exceptions, all the classes of the animal kingdom, and if their number is great among fishes, they are not less numerous in mammals and birds. The higher classes of animals usually inoculate themselves through the intermediation of molluscs, worms, and crustaceans, and it is therefore in the ranks of these that we must seek for their first abode. Without admitting that their size bears some proportion to the host which gives them shelter, still, the largest species, the *Distomum Goliath*, is found in the liver of one of the balænoptera. This distome is of the size of a large leech, and its host does not measure less than twenty metres.

Mons. Willemoes-Suhm mentions a distome which at the time of its cercarian evolution lives freely in the water, and attaches itself by its sucker to the larvæ of worms or copepod crustaceans, and then lodges in their dejecta without encysting itself. This is the *Distomum ocreatum* of the herring, according to Professor Moebius. Mons. Ulialnin found in the bay of Naples another free distome, which is also attached by its ventral sucker to certain copepods, and which becomes the *Distomum ventricosum* inhabiting many kinds of fish.

Any one who wishes to make observations on distomes in the state of cercariæ has only to examine some
fresh-water molluscs, either the Limneæ or Planorbes
found in ponds ; as he tears the animal to pieces on the
stage of a simple microscope, he will not fail to perceive
a multitude of struggling and wriggling tadpoles.   Their
tails twist with each other, furl up, extend, and describe
arcs of circles, as if we had a nest of serpents under our
eyes.

Each species of distome has it own cercariæ, which
are scattered among as many
different inferior animals. Birds
and fishes become infested by
them in consequence of eating
these animals.

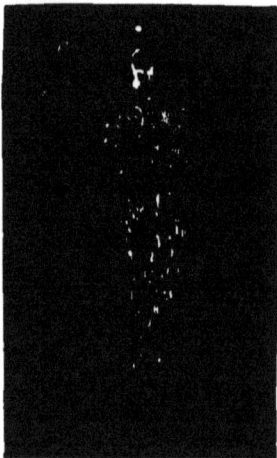

We may here cite as an
example of this class of parasites the *Distomum hepaticum*,
or liver fluke ; this species is the
most interesting to us of all the
genus ; it attains the size of a
moderate leech, and habitually
resides in the liver of the sheep.

Fig. 42.—Liver fluke of twice the In order to discover it, we have
natural size ; *a*, mouth ; *b*, penis;
c, digestive tube; *d*, abdominal only to examine a fresh liver.
sucker.

They are usually found in the
biliary canals, where they move about like planariæ.   It
is always of a deep colour, and is doubtless introduced in
the state of cercaria, when the animal is drinking.   M.
Willemoes-Suhm supposes that the *Distomum hepaticum*
has for a vehicle a small snail, the *Limax agrestis*,
which the sheep swallows with the grass on which it

feeds. Its principal abode is in the ruminants and only casually in man. It is said to be unknown in Iceland. The *Distomum lanceolatum* has also been found in man.

Dr. Bilharz, the pupil of Siebold, discovered in the year 1851, on man, a parasite in every respect remarkable. It belongs to the family of the Distomidæ, and on account of its peculiarities, it has been made into a genus under the name of *Bilharzia*. It is found in Egypt, and lives in the vena portæ and in all its ramifications in man. According to Bilharz, this distomian is diœcious, the male being of considerable size, the female slender and delicate, which fact does not agree with the usual characteristics of diœcious animals. At least half of the Fellahs and Copts suffer from these parasites; these worms, at the period when they lay their eggs, proceed from the vena cava to the veins of the pelvis, and after having produced very grave consequences, they are at last evacuated with the urine.

Another distome was also found by Bilharz in the intestines of a young Egyptian boy.

The largest known distome inhabits the liver of the *Balenoptera rostrata*, the little whale of thirty feet in length, which is regularly met with on the coast of Norway. The intestines of the ordinary seal often contain a very curious distome, which was first observed by Rudolphie, the *D. acanthoides*. The seal is also infested by the *Distomum cornus*, which some have incorrectly preferred to place in the genus Amphistoma.

Besides the distomes which inhabit the liver, there are found but few in the mammalia, except in the Cheiroptera: these insectivorous animals have their intestines literally full of these parasites. We have noticed the species which regularly frequent our bats, and it only remains to discover the insects by means of which they are introduced; for it is probable that these insects are infested by cercariæ during the time that they inhabit the water. Larvæ and their parasites ought to be carefully studied in the localities where bats abound.

There are few birds, especially among the grallæ and the palmipedes, which do not enclose in their intestines a certain number of distomes. The same may almost be said of reptiles and batrachians, but it is especially in fishes that their number is greatly increased. We may say that there is no fish which does not nourish some of these trematodes. Among a portion of these, the cycle of evolution and transmigration is perfectly known; we may instance the *Distomum nodulosum.* This worm inhabits the intestines of the perch.

The scolex, as well as the cercaria, has its particular characters, and we have long since found the latter in a fresh-water mollusc, the *Paludina impura.* The cercaria is easily recognized by the presence of two particular folds at the base of the buccal bulb, and by the transparency and the form of the extremity of the urinary apparatus. In the adult distome, this same part of the urinary apparatus encloses large vesicles with very distinct partitions.

We may also mention among the distomes a species from fish, which has a great affinity with the singular distome observed by Bilharz, of which we have spoken above. This distome inhabits the "castagnole," or *Brama raii.* Under the opercula of this fish, the skin is folded, and forms one or more pouches, in each of which lives a coupled distome, that is to say, by the side of each large and fat individual, full of eggs, there is one which is slender.   It is the *Distomum filicolle*, to which the name of *Monostomum* was at first given.   We should be correct in supposing that of these two hermaphrodite worms one acts rather as a female, the other as a male. It is doubtless in this sense that Steenstrup maintained his assertion, that there are in nature no hermaphrodites.

Thus there are two kinds of distomes: the first live in couples in a cyst, the second in couples joined together, but at liberty; and in each case only one individual produces eggs.   These are distomes which act really like diœcious worms.   We find, however, a more remarkable instance in the *Monostomum bijugum* of Miescher.   In the tumours which are formed in the beak of the grosbeak (*Fringilla*), he has constantly found two individuals; and in many cases he has surprised them with the penis of one engaged in the sexual organ of its companion.   These worms, while they live in couples, resemble each other like snails and leeches; they are mutually fecundated, and both lay eggs.

Leuckart recognized these sexual distomes in their cyst, in the larvæ of ephemerides; and Linstow noticed a distome thus sexual and encysted in the *Gammarus pulex*.

The name of Monostoma has been given to some of these trematodes which have no abdominal sucker.

One of the most curious worms of this group is the *Monostomum mutabile*.   It lives in the sub-orbitary sinus of several aquatic birds; that is to say, in the nasal fossæ, especially of water-rails and moorhens.   We give a slightly magnified representation of them.   It is a worm resembling an elongated leaf.   By compressing

it slightly on the stage of the microscope, we easily discover the ovary, the matrix, and oviduct full of eggs.
By isolating some of the eggs, and crushing them gently
to break the shell, we set free the worm (Fig. 44), quite
different from the mother (Fig. 43). The former has
two eyes surrounded by a ciliated mantle, and by means
of this ciliated envelope, the monostome swims freely in

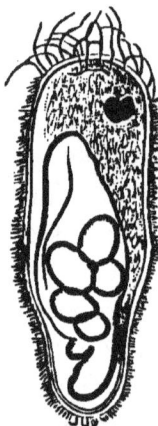

Fig. 43.—Monostomum   Fig. 44.—Monostomum mutabile. Ciliated embryo with
mutabile (adult).   sporocyst and young cercariæ, greatly magnified.

the water. If we compress it slightly, we see that in the
interior of the ciliated covering, there is still another
animal, without eyes, without ciliæ, and of an entirely
different form, which in its turn encloses a whole progeny.

The embryo, having long ciliæ in front, and in the
interior a sporocyst already full of young cercariæ, is
shown in Fig. 44. It is this latter creature which the
ciliated embryo must confide to the care of others; this
she puts out to nurse with some mollusc or other, until
it is fit to provide for itself in its turn. We have still to
discover the train by which the parasite must travel, in

order to arrive again at the nasal fossæ which are the first cradle of the family.

We find occasionally between the feathers of some birds tubercles of the size of a pea, and when we open them we see in each two similar worms, placed so that the stomach of one is applied to that of the other; this is the monostome of which we have spoken above. These worms are from three to four millimètres in length (about ·13 in.), and are found in the titmouse, the siskin, the sparrow, the canary, and some other birds.

A worm very common in the intestines of the green frog is known by the name of *Amphistomum sub-clavatum.* Its cercariæ are usually found in an acephalous mollusc, known by the name of *Cyclas cornea.* That which

Fig. 45.—Cercaria of *Amphistomum sub-clavatum.*

Fig. 46.—Sporocyst of *Amphistomum sub-clavatum* from the *Cyclas cornea.*

distinguishes the scolices of this species is the great contractibility of the external membranes of the young individuals; they lengthen, they shorten, they swing to the right and the left, describing a semicircle on the anterior half of the body (Fig. 46). We represent side

by side the cercaria of this amphistome, and the adult
and sexual amphistome, as it is found in the intestines
of the frog.

Constantine Blumberg has recently published an
interesting memoir on the structure of the *Amphistomum
conicum.*

A beautiful trematode worm, known by the name
of *Hemistomum alatum*, whose antecedents have not been
ascertained, lives usually in the intestines of the fox. It
is about four or five millimètres in length (about ·17in.).
Many birds harbour Holostomes which belong to the
same group, the first state of which is not yet known.
The *Holostomum macrocephalum* is common in the intestines of rapacious birds; it is from five to séven millimètres in length (about ·23 of an inch).

We close the history of trematode worms by giving
the figure of a beautiful one known under the name
of *Polystomum*, which lives in its adult state in the
bladders of frogs (Fig. 48).   Interesting observations
have recently been made on the manner in which they
are introduced into the bladder.

The worms which naturalists call Cestoïds, or Cestodes
(which means, like ribbon or tape), have for their type
the tape-worm known by every one. They are very
abundant in many animals, are found in almost every
class of the animal kingdom, and are almost as common
as the distomians, of which we have just spoken.   They
are introduced into animals which are vegetable-feeders,
by means of water and plants, and into carnivorous
animals by their prey. The tape-worms of the herbivora
lay eggs like the others, but their embryos have, as soon
as they are hatched, a ciliary covering which allows them

to live and move about in the water.   Those of beasts of
prey are entirely different; it is by means of the prey
that they enter their hosts.   Each carnivore has its own
worms, as it has its own prey which introduces them.

Fig. 47.—*Amphistomum sub-
clavatum* of the frog.

Fig. 48.—*Polystomum
integerrimum.*

Independently of these worms, the vegetable-feeders
afford lodging to some which are not their own.

We have found in bats two tæniæ, both incompletely developed, and occupying the digestive tube. One has a rostellum without hooks, like the tæniæ of the vegetable-feeders, the other has hooks like those of the carnivora. These cestode parasites are observed to be of two principal forms ; the first vesicular, like the finger of a glove partly drawn inwards. They are always lodged in the midst of the flesh, or in a closed organ in the middle of a cyst; under this form the cestode worm is harboured by a host which is to serve as a vehicle to introduce him into his final host. He is a parasite on a journey ; he is always agamous, and usually bears the name of cysticercus (Fig. 49). As to the second form, it is like a ribbon ; it attains a great length, always occupies the intestine, attains its complete and sexual development, and lays an innumerable quantity of eggs which are disseminated with the evacuations.

Fig. 49.—Cysticercus ; *a*, upper part of the vesicle ; *b*, place where the vesicle is about to separate ; *c*, neck of the worm ; *d*, the head, showing the suckers and the crown of hooks.

The rabbit harbours a cysticercus which has its final destination in the dog (a xenosite); but independently of this stranger, it gives hospitality to a special tænia in its intestines. This is its own worm, the *Tænia pectinata*, which is a nostosite. All the herbivora are in a similar case; the ox and the sheep possess a peculiar tænia of their own, besides those which they lodge for the sake of the carnivora. The worms of the herbivora have particular characters by which they are easily known; they have no crown of hooks.

The tænia of the wolf, which has often been confounded with the *Tænia serrata*, lives in the brain of the sheep, and produces a disease known as the "gid." It was formerly said that every animal has its enemy. We should rather say that each species has its parasites, and each parasite has its vehicle by which it is introduced.

These tape-worms are found in all the vertebrate classes. An herbivorous animal usually serves as a vehicle, but it more frequently carries, besides its passengers, species which are peculiar to itself. As the carnivorous animal is not intended to be eaten like the herbivora, it cannot serve as a vehicle, and if by chance its muscles enclose some passenger, he has lost his way and that for ever.

Do the cetacea generally live on fish, and do they become the prey of some aquatic carnivora? We have reason to think so, from the presence of certain agamous cestodes, which have been frequently found in too great number to allow us to suppose that they have lost their way in these aquatic mammals. There have been seen in the substance of the muscles of many species, or rather in the layer of blubber which covers the skin, agamous cestode worms of the genus *Phyllobothrium*, which can only accomplish their evolution in some large *squalus*. There must then be contests between dolphins and sharks, contests in which the dolphins are worsted, in spite of their superiority. These Phyllobothria have been found in the *Delphinus delphis*, the *Tursio*, and the *Ziphius*. As the *Orca* attacks the whale, and feeds upon its flesh, there would be nothing surprising in our finding in these large cetacea, some agamous cestode destined to pass through the last phase of its evolution in this terrible carnivorous animal.

The cestode can scarcely be called a parasite under the first vesicular form. It is sufficient for it to pass through its first transformation in the midst of the tissues, and it will remain weeks, months, even years, without undergoing any change; it asks for nothing but an hospitable roof; and this mysterious being, that had often come they knew not whence, encamping rather than lodging, always without progeny, was long since cited by the naturalists of a former age in favour of the old hypothesis of spontaneous generation.

It is not the same with the second form. Here the worm, always lodged in the intestines, grows with extraordinary rapidity, and fulfils all the conditions of a true parasite. In a fertile soil it extends itself and produces young as long as it has any life, and in no group of the animal kingdom do we find any fecundity to be compared to that of this worm. Boerhaave described a broad tapeworm, three hundred ells in length. Eschricht estimates the number of the segments of this worm as ten thousand; and if we consider that each segment, or, we should rather say, each complete worm, may perhaps enclose thousands of eggs, we may form some idea of the profusion of germs which can be scattered by each individual.

To thoroughly know an animal we must have made observations on it during all the phases of its evolution.

Let us sketch these phases. All the cestodes have eggs, usually in great number, very well protected against external agents. They endure heat and cold, drought as well as humidity, resist by means of their envelopes the most violent chemical agents, preserve the faculty of germinating, we will not say for weeks, months, and years, but for centuries. When they first leave the egg, we see an embryo of an oval form, transparent, composed apparently of sarcode, contractile throughout all its extent, and in the middle of which we perceive six stylets arranged in pairs, and which at last move with great rapidity.

The following is the manner in which, some years since, we described these six hooked embryos produced by a tænia of the frog, which were struggling by the side of each other on the slide of a microscope. "The six hooks are arranged regularly in each individual, and move exactly in the same manner. They are very slight, and of nearly half the diameter of the embryo. Two occupy the median line, and unite like a single stylet; these are nearly straight, and a little longer than the others. They only move backwards and forwards. Their action is like that of the parts of the mouth in certain parasitical crustaceans, the Arguli, when they endeavour to pierce through the tissues. They are in continual motion to and fro. The other four hooks are similar to each other, and differ from the first in the point, which is curved into real hooks. They are arranged two and two, to the right and left of the first, so that they all meet at the base. Their movements are not the same as those of the two first; they remain almost fixed at the base, while they describe a quarter of a circle at the extremity. Let us imagine the six hooks, placed in front in the same direction. The two in the centre advance, and the two pairs placed symmetrically by the side of them, are lowered and drawn backwards, and thus push the body forwards.

"It is like the dial-plate of a clock, with three hands placed by the side of each other; that in the middle would advance directly forward, while the two others would be lowered until they formed a right angle with the first. This is the movement which we observe in all the stylets. The result of this is that we distinctly see the embryo penetrate between the *débris*, or into the crushed tissues which surround it. These embryos imitate the movements of a man who wishes to get through a window a little above him, and who, having succeeded in passing his elbows through, pushes his body forward by leaning them on the frame.

"We see the same efforts continue for hours; and we can easily understand that there is no living tissue, however dense it might be, except the bones, which could not be easily penetrated by these microscopic embryos. This explains why we so commonly find cysticerci scattered in cysts along the intestines and between the membranes of the mesentery, and how they can, by piercing the walls of the vessels, spread themselves into the most distant organs, by means of the blood which conveys them. When the embryos have once pierced these walls, they hollow out the tissues in all directions, until they find themselves in the muscles, or in the organ which is indicated in their itinerary. When they have arrived at their destination, they stop and surround themselves with a sheath; their stylets, which are no longer of use to them, decay; and at one of the extremities appears a crown of new hooks quite different from the former ones, which will serve to anchor their progeny in the new host into which they may be introduced."

Thus the vesicular worm (Fig. 50), fully formed, and without undergoing any change, waits till its host, or the organ which shelters it, is eaten, and then wakes up in the stomach. Every living cysticercus which penetrates into the stomach, instantly quits its torpid state; it gets rid of its useless parts, abandons its former cavity, penetrates into the intestine, attaches itself by its new hooks and its suckers to the enclosing membranes, and grows with such rapidity, that in less than six weeks, we often find a tape-worm many metres in length. The vesicle which had hitherto protected it is abandoned, and the part which remains with hooks and sucker is the mother which has produced in this agamous manner the whole colony. This mother is usually called the head of the tænia, or more properly the scolex. As long as the mother is there, she engenders and produces cucumerinæ, that is to say, proglottides, which are the perfect and sexual state of the cestode.

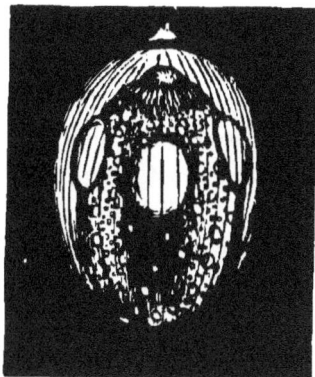

Fig. 50.—Vesicular worm.

We have seen among the trematodes a worm of a particular form leave the egg, and immediately produce a swarm of young ones, which go and live

11

separately. In the cestodes all these individuals are united in a kind of band, and are besides this joined to the mother, which becomes the root of the family. This root, planted in the walls of the intestine, is the head. Thus each segment of the tænia is an individual, and at the period of sexual maturity, this individual is detached, goes away with the feces, spreads over the grass or elsewhere, and thus sows far and wide the eggs which it contains.

The tænia, as well as the other tape-worms, is generally looked upon as an imprisoned parasite during the whole of its existence. This is a mistake; the last stage of the life of cestodes is a phase of liberty. The cucumerina, or, as we have proposed to call it, the proglottis, that is to say, the complete and sexual animal, is evacuated with the feces; and when we notice a dog leaving his dung upon the grass, it is not uncommon to see there worms which move like leeches, and whose white colour is in strong contrast with the mass which contains them. The duration of this last stage is very short, it is true; but it is, nevertheless, during this period of her life that the mother scatters the eggs which are to disseminate the species.

We repeat that each animal has its parasites, and these in their turn are not always exempt from them. We have already cited some examples of this.

Man has the dental system of a vegetable feeder; but, thanks to fire, which he alone knows how to produce and maintain, he eats flesh. It is by these means that he nourishes the solitary worm, which, by its crown of hooks, is a cestode belonging to the carnivora, and the *Tænia mediocanellata* with the *Botriocephalus*, which are cestodes peculiar to vegetable-feeders. As a feeder on vegetable diet he also harbours vesicular agamous cestodes, which are only found in him as passengers.

The *Tænia serrata* of the dog lives at first as a passenger in the peritoneum of the hare and the rabbit; and every one knows how greedily the dogs eat the viscera of these animals.

The cat entertains another kind of tænia, and, as we may easily suppose, in its young state it lives as a passenger in the mouse or the rat. Who then has traced out for it this itinerary, and pointed out the way, the only one by which the parasite can hope to take possession of its proper abode? Evidently it is neither the tape-worm nor the cat. The plan for all these various species is marked out beforehand, and each animal as soon as it is born knows it without being taught.

A Danish naturalist, Mons. H. Krabbe, has just finished a special work on cestode worms of the genus *Tænia*, and he remarks that there is no class in which these worms are so abundant as in that of birds. It is among the rapacious and carnivorous birds of this class that they are less abundant. Among mammals, the carnivora possess the greater number. This fact, as M. Krabbe remarks very rightly, seems to indicate that the cestodes of birds especially employ the inferior aquatic animals as their vehicles when in their incomplete state.

Let us consider the solitary worm of man (*Tænia solium*), it will enable us to understand all the others. Known by the name of tænia, or solitary worm, it is, like all the cestodes, a marvellous association of mothers and daughters, which are developed and vegetate in

a peaceable community. Each segment is a complete being, which encloses within itself an entire and very complicated apparatus for the fabrication of eggs.

We give (Figs. 51 and 52) the representation of a solitary worm, peculiar to man, of the natural size; and at the side the scolex, usually called the head, slightly magnified.

Under its first vesicular form the solitary worm is

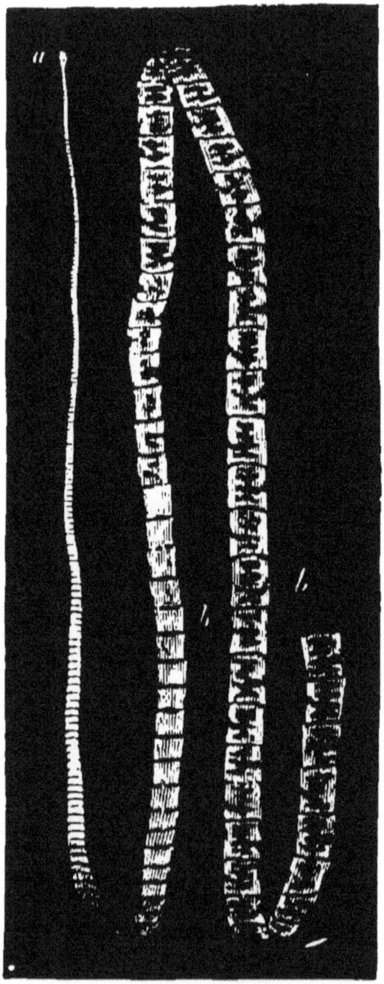

Fig. 51.—*Tænia solium*, or solitary worm ; *a*, head, or scolex ; *b*, tape formed of many individuals, the last of which, completely sexual, separate under the name of *proglottides*, and represent the adult and complete animal. Each solitary worm is a colony.

Fig. 52.—*a*, Rostellum ; *b*, crown of hooks ; *c c*, suckers ; 1, scolex of the tænia solium ; 2, hooks expanded ; *a*, heel of the hook.

planted in a provisional soil. After this it is transplanted into a richer soil, where it flowers and throws out its numerous seeds. It comes to us from the flesh of the pig, in which there lived vesicular worms, of the size of a hazel-nut. The muscles are sometimes full of them, and the pig is then said to be " measly." The ancients noticed that the sucking-pig never takes this disease ; and as *Sus scropha* is the name of the pig, the term scrophula has the same origin as the specific name proposed by Linnæus.

The measles in pork have been attributed to damp, to feeding on acorns, to hereditary causes, to contagion, even to injured corn and mouldy bread, All these theories we find in pathological treatises. The only true cause, however, is the introduction of the eggs of the *Tænia solium* into the intestines. If we wish to prevent this infection, we must not permit the animal to eat man's excrements, nor to drink water in which substances that have become decomposed on a dung-heap have been allowed to remain.

The cysticercus of the pig, when introduced into man, becomes a tænia with as great certainty as the seed of a carrot will produce this plant if sowed in suitable soil. The observation had been for a long time made without any explanation being given, that this parasite especially shows itself among pork butchers and cooks. This is because these persons, more frequently than others, handle raw pork. The same observation has been made respecting children who have made use of the gravy of raw meat. Minced raw meat (*conserve de Damas*) has been prescribed with success in chronic diarrhœa. The tape-worm has often been known to make its appearance after this treatment, as may well be supposed. Tænia helminthosis is constant and general in Abyssinia, and they there commonly eat raw beef. Those who do not eat meat, as the monks of certain orders there, who live only on fish and flour, never have the tænia. Ruppell and many others have noticed this fact. Mons. Küchenmeister says that at Nordhausen, in the Hartz, as well as throughout all Thuringia, measles are very prevalent among pigs; and as the people are in the habit of eating minced pork, both raw and cooked, spread on bread for breakfast, this country may be looked upon as the Abyssinia of the north.

The doctor at Zittau caused a man who was condemned to death, to take, seventy-two hours before his execution, some cellular cysticerci from a measled pig; and he found in the duodenum of the man four young tæniæ, and six others in the water in which they had washed the intestines. The latter had no hooks, but those of the former had some in every respect similar to those of the *Tænia solium.*

We have ourselves caused a pig to swallow eggs of the tænia, and have given it the measles. Messrs. Küchenmeister and Haubner, who were ordered by the government of Saxony to make some experiments, also caused three pigs to swallow eggs of the *Tænia solium*, and two of these were affected with measles. A piece of flesh, weighing 4½ drams, contained 133 cysticerci, which amounts, for 22 German lbs., to 88,000 cysticerci.

The use of raw pork will produce tæniæ more readily than raw beef. Dr. Mesbach has given the following instance in support of this fact. At Dresden, a father and his children regularly ate, at their second breakfast, raw beef, but one day they took pork instead, and eight weeks afterwards one of the children, when in the bath, voided two ells of *Tænia solium*.

The etiology and prophylaxis of the solitary worm, that is to say, its mode of introduction, and the means of protecting ourselves from it, are clearly indicated. It is sufficient to introduce one of these vesicles into the stomach in order to have the tape-worm. The experiment has been made : young men have ventured, in the interests of science, to swallow some, and have ascertained how many days were required for the parasite to be sufficiently complete to give off segments with the feces.

These vesicles in pork come from the eggs which the tænia has scattered in its passage, and if the pig comes by chance in contact with the fecal matter of a person infested by one of these worms, it is soon infested and becomes what is called measled ; in this fecal matter there are either free eggs which have been evacuated by the worm, or else fragments, known long since under the name of cucumerinæ, which are full of eggs.

These fragments of tænia, which I have proposed to name proglottides, and which are nothing else than the worm in all its sexual maturity, are still living and wriggling at the moment of their evacuation, or else they are dead and often completely dried ; but in either case, they are full of eggs. Each egg is surrounded by membranes and shells, which effectually protect it against all dangerous contact.

A fragment of the mature tænia, thus filled with eggs, when introduced into the stomach of the pig, is rapidly digested, and the eggs are set at liberty. These lose their shells by the action of the gastric juice, and there issues an embryo singularly armed. As we have before said, it carries in front two stylets in the axis of the body, and on the right and left sides two other stylets curved at the end, which act like fins. These embryos bore into the tissues as the mole burrows into the soil. The middle stylets are pushed forward like the snout of the insectivore, and the two lateral stylets act like the limbs, taking hold of the tissues and forcing the head forwards. In this manner the embryos perforate the walls of the digestive tube.

An egg of the *Tænia solium* may be swallowed by a man instead of passing into the stomach of the pig. It is hatched in his stomach precisely in the same manner, and the embryo takes up its lodging in some enclosed cavity. Some have been found in the eye-ball, in the lobes of the brain, in the heart, or in the muscles. We have lately read an account of the effects produced by one of these wandering worms, on a man who died after suffering from a peculiar disturbance of the mind. Two spirits seemed to haunt and speak to him, the one a German, the other a Pole. Filthy images were called up before his imagination. At the post-mortem examination, cysticerci were found to occupy the sella turcica, near the commissure of the optic nerves. One of these was alive, the others were calcified. Two others in a similar condition occupied a lobe of the brain.

Man harbours not only the *Tænia solium*, but another species very similar, which naturalists have only learned to distinguish from it during the last few years, the *Tænia medio-canellata*. We give a magnified representation of the scolex, that is to say, of the head of this worm, which has no crown of hooks in the middle of its four suckers.

This solitary worm is introduced by means of beef, and the cysticercus, during its abode in the cow, manifests already the peculiar characteristics which enable us to recognize the species, that is to say, no crown of hooks, but four suckers, and in the middle of them, some blotches of pigment. Leuckart fed a calf with eggs of this tænia, and at the end of seventeen days, the animal died of acute miliary tuberculosis, produced

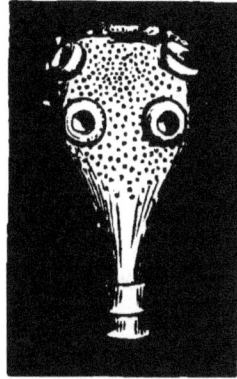

Fig. 53.—Tænia medio-canellata.

by the great abundance of cysticerci. This second species, which had been always confounded with the preceding, and which is nevertheless the more common, has therefore a different origin from the *Tænia solium*. Observations made quite recently in the north of Africa demonstrate this. Great difficulty had sometimes been felt in explaining the presence of the tænia in persons who had not eaten pork. This embarrassment arose from the confusion of the two species, and this confusion is the more easy as the head of the colony must necessarily be found in order to distinguish them.

Scharlau, at Stettin, found tæniæ in seven children who had been fed, on account of anæmia, with raw meat. The tæniæ were those of this species. We have ourselves found them in children to whom the use of raw meat had been prescribed.

We do not think it necessary to speak here of a third species of tænia (*T. nana*), which also lives at our expense, but which has been hitherto found only in Egypt.

We know perfectly well the itinerary of the *Tænia serrata* of the dog, which is so abundant, that there are few of these animals that do not enclose some and even many of them. There are few except lapdogs which do not harbour them. We can easily assign the reason. Every tænia, like every animal, has its eggs ; each plant has its seeds. These eggs are laid by the mother in the most favourable condition for the development of her progeny. The dog deposits its dung on the grass rather than in any other spot, because the eggs of its tænia, which are destined to the rabbits or hares, will have greater chance of arriving at their destination than if they were exposed on the bare earth, or in the water. Their prodigious number is calculated according to the chances of their arriving safely. The egg, when introduced into the stomach of the rabbit, is rapidly hatched in this organ under the action of the gastric juice, and the embryo which is produced from it seeks its hiding-place in the midst of the tissues which surround it ; it bores into them, and establishes itself in the folds of the peritoneum. Then, once in its resting-place, it barricades itself, and waits patiently for an opportunity of introducing itself into the stomach of the dog.

This microscopic embryo is armed with six hooks, like embryos of all the cestodes ; it employs them with much dexterity to pierce the walls of the organs, and to hollow out a space for itself in the substance of the tissues. Shut up in its hiding-place, membranes form around for its protection ; its six hooks, having become useless, wither ;

other hooks in the form of a crown appear by the side of four rounded projections, the future suckers; and, sheathed in a large vesicle full of a limpid fluid, it waits patiently for the moment when it will find a place in the stomach of a dog. If good fortune awaits it, it will wake up, some fine day, in the stomach of the animal which has eaten the rabbit, its former home, and a new life will commence for it. The organs in which it was imprisoned are digested, it gets rid of all its swaddling-clothes, unrolls itself, separates from the vesicle which has protected it hitherto, and penetrates into the intestine; there, immersed in the food of its host, it grows with extreme rapidity, and assumes the form of a ribbon or tape. The ends of this tape are successively matured, detach themselves, and become the complete worms, full of eggs, which are evacuated with the feces; scarcely have they made their appearance in the open air before they burst and scatter their eggs.

He whose scientific curiosity is sharpened, has only to watch the dung of the dog at the moment of its evacuation to distinguish on its surface worms of a milky-white colour, contracting like leeches, which are the true *Tænia serrata* in its adult state. Experiments made on this species have given sanction to what I had said respecting the cestodes.

The tænia, under the name of *Cysticercus cellulosus*, lives in the folds of the peritoneum of the rabbit and the hare, and passes directly from the rabbit to the dog to become complete.

It is very curious that the fox, so nearly allied to the dog in appearance, and which also eats rabbits, never has the *Tænia serrata*, but this animal nourishes other worms.

It was with these cysticerci that I made experiments on four dogs, which I took with me to Paris, in order to convince those who could not believe in the migration of parasites. It was this species that I gave also to the dogs which served as a demonstration at Paris at the course of lectures given by Mons. Lacaze Duthiers.

Some years ago, while making a post-mortem examination, at the Museum of Paris, of some young dogs which I had previously infected with *Tænia serrata* at Louvain, there were found by the side of these some *Tæniæ cucumerinæ*. These dogs had taken nothing but milk and cysticerci! Whence came these *Tæniæ cucumerinæ?* I knew not, and I frankly owned it to the members of the Commission who proposed the question to me. This however did not prevent my being greatly puzzled with the presence of this worm of whose origin I had no idea. Now we know whence they came. An acaris, the Trichodectes, lives in the hair of young dogs and harbours the scolex of this cestode. The dog, by licking its own hair, grows infested, like the horse, which in a similar manner introduces the gad-fly, and although it has taken no other nourishment, harbours its own epizoaria.

The name of *Cysticercus tenuicollis* has been given to a vesicular worm which inhabits the peritoneum of the ox, the goat, the sheep, &c., and which turns to a tænia in the digestive tube of the dog. Mons. Baillet has made the principal experiments on this transmigration. The itinerary of another cestode worm, the *Cœnurus* of the sheep, is to pass through the sheep in order to reach the wolf or the dog. This worm has only lately been recognized in its tænoïd form; it has, on the contrary, been long known under the name of *Cœnurus cerebralis;* this develops itself on the brain of the sheep, and occasions the disease known by the name of "gid." This disease may be produced artificially. The sheep which swallows the eggs of this tænia shows the first symptoms of it towards the seventeenth day. If we kill it at this time, we find on the surface of the brain, either at the base or the summit, or sometimes between the hemispheres and the cerebellum, one or more white vesicles of the size of a pea, and on which no traces of buds are yet to be seen. This vesicle, of a milky-white colour, and filled with liquid, is the scolex. Near these vesicles are to be seen some very irregular yellow furrows, like tubes abandoned by some tubicolar annelid; this is the gallery through which the vesicular worm has proceeded to the place where it has been found.

A fortnight later, that is to say, about the thirty-second day, the cœnurus is as large as a small nut, and one can see with the naked eye some small nebulous corpuscles, separate from each other, of the same form and size; these are the buds or scolices which have risen up, but which, as yet, have neither hooks nor suckers.

We give the representation of one of these vesicles, on the internal walls of which

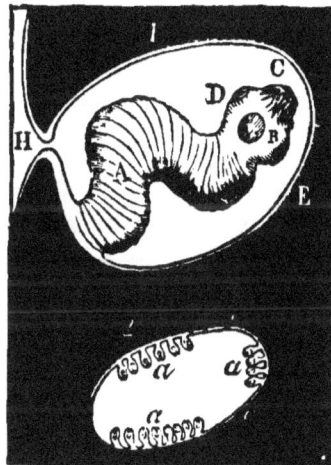

Fig. 54.—Cœnurus of the sheep. · 1, the enclosed scolex ; 2, Hydatic vesicle, with the scolices in their place within it.

young scolices have been developed; this is nearly of the natural size. Fig. 2, *a*, *a*, shows these scolices of nearly the natural size. Fig. 1 represents an isolated and magnified scolex ; A, shows the segments of the future proglottides ; D, the suckers ; C, the hooks ; H, the vesicle which contains them.

Eggs of the same tænia have been given to sheep at Copenhagen and at Giessen, and Messrs. Eschricht and R. Leuckart have obtained the same result as we had at Louvain. On the fifteenth or sixteenth day the first symptoms of "gid" declared themselves. At about the thirty-eighth day the crown of hooks appeared, the suckers were formed, and the whole head of the scolex was sketched out. All these heads can leave or enter the sheath at the will of the animal. It is truly a polycephalous animal when the scolices are expanded. This worm continues to grow for a long time in the cranial cavity, and produces by its presence the gravest results. The sheep necessarily dies at last, unless we remove the parasite by means of the trepan.

The cœnurus, at this point of development, swallowed by a dog, undergoes great changes in a few hours. The proscolex, or large vesicle, withers ; the different scolices unsheath their cephalic extremity, become free, penetrate into the intestine with the food, and attach themselves to its walls, so as to form as many colonies of tænia as there are distinct heads. A dog which has swallowed a single cœnurus may therefore contain a considerable number of tæniæ.

The development of this worm proceeds very rapidly, and it only requires three or four weeks to attain many feet in length. The organization of this worm, in the state of strobila and of proglottis, is in every respect like that of the *Tænia serrata;* we have even endeavoured in vain to distinguish these worms from each other by their hooks. The wolf or the dog follows the flock of sheep, scatters the proglottides or the eggs in their way, and the sheep, browsing on the grass with the eggs attached, become infested with their most dangerous enemy.

To arrest this disease, only one thing is necessary, to destroy by fire the head of every sheep attacked by the "gid." The rest of the animal may be eaten without danger.

Pouchet did not succeed in giving sheep the "gid" at first, for the very simple reason that he employed the eggs of the *Tænia serrata*, instead of those of the *Tænia cœnurus*; he had confounded the two species. The cœnurus of the sheep is a true calamity when it spreads in a country. The animal attacked by it is lost, and the mischief may be indefinitely propagated by giving as food to dogs the head of the sick animal, with thousands of young tæniæ enclosed within each.

There exists a singular cestode which bears the name of *Echinococcus*. We give a figure of the echinococcus of the pig, slightly magnified, and an isolated scolex (Figs. 55 and 56). In its first form it is composed of closed sacs, which grow to the size of a nut, and sometimes to that of an orange. It usually lodges in the liver of the pig, but establishes itself also in man. We have been assured that part of the population of Iceland have been attacked by it. The abundance of this parasite in that country is attributed to the want of cleanliness, and the number of dogs that they keep around them. The echinococcus becomes a tænia in this animal. It scatters the eggs with its dung, leaving them directly or indirectly on plants which the Icelanders eat ; for they gather for food certain mosses, sorrel, cochlearia, dandelion, &c., from the midst of the plains in which live flocks of sheep guarded by dogs.   The eggs are scattered everywhere on plants or in the water.

Leuckart has made some very interesting experiments on the echinococci.   In Fig. 57 is shown a tænia which proceeds from an echinococcus.

Fig. 55. - Isolated scolex of the *Tænia echinococcus* from the pig.

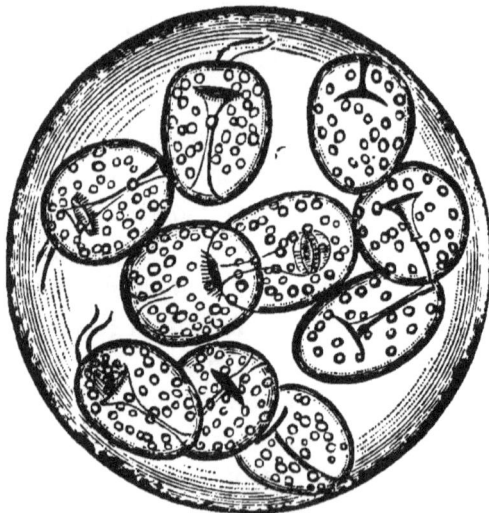

Fig. 56.—*Tænia echinococcus*, from the pig.

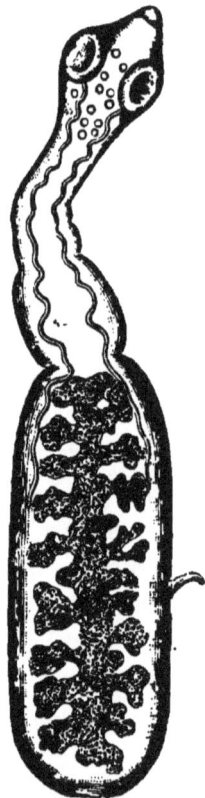

Fig. 57.- -*Tænia echinococcus*, from the dog.

There is yet another tape-worm harboured by man, the *Tænia lata*, better known under the name of Bothrio-

oephalus. We give in Figs. 58, 59, and 60 representations
of this worm in the state of a colony, also the scolex or

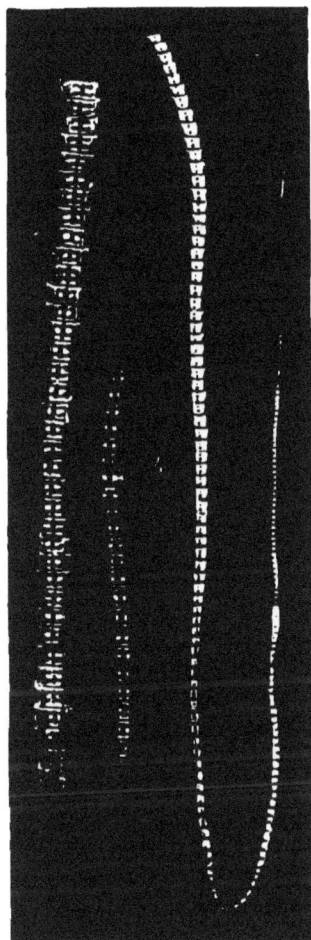

Fig. 59. -Bothriocephalus latus, scolex.

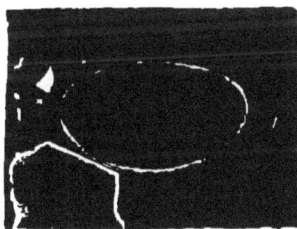

Fig. 58.—Bothriocephalus latus. *a*, scolex,
b, the proglottides, *c*, the sexual organs.

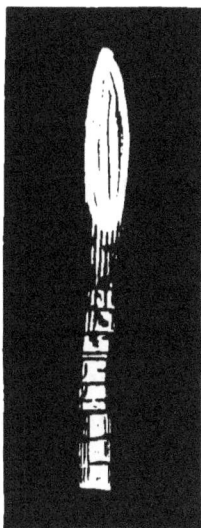

Fig. 60.—Bothriocephalus latus,
egg.

head separately, and an egg. Its history is very curious,
especially with reference to its geographical distribution.

It is only found in Russia, Poland, and Switzerland, and the limits of the places which it inhabits are perfectly defined. Siebold, during his stay at Königsberg, could determine from the nature of the worms, whether the patient who consulted him lived on one side or the other of the Vistula.

A Russian naturalist, Dr. Koch, thoroughly studied this interesting worm and its evolution. He says that this cestode is rare at Moscow, while at St. Petersburg, Riga, or Dorpat it is common. If this be really the case, it must doubtless be attributed to the fact that in one place the inhabitants drink spring water, and in the other water from the river.

A very curious circumstance is the actual rarity of the Bothriocephalus among the inhabitants of the shores of the Lake of Geneva, though formerly it was very common there. This diminution, if we may not call it disappearance, is due to the change which has been made in the construction of water-closets, all of which formerly emptied themselves into the lake, so that the embryos were hatched in the water, and persons were infested by them through drinking it. At present the refuse of the towns is carefully collected for the purpose of manuring the land. This is the result of the advice of Mons. de Candolle, half a century ago; for this naturalist clearly understood how great was the loss to agriculture from the neglect of this fertilizing agent.

The itinerary of this tape-worm is simple. It passes from man to the water under the form of an egg, or of a proglottis ; and from the water to man in the shape of a ciliated embryo. In this manner it is introduced with the water that is drunk. The Bothriocephalus, like other cestodes, is free at the commencement and the end of its life : at the beginning, in order to penetrate into its host ; at the end, to scatter its eggs.

Messrs. Sommer and Landois published, in 1872, an anatomical description of the sexual organs of the *Bothriocephalus latus*, of such completeness, that it will be long before any one will again take up this sub-ject, which had so much occupied helminthologists ever since the celebrated work of Eschricht. This memoir is illustrated by superb engravings, which represent these organs under every aspect. Dr. Böttcher, of Dorpat, found in the small intestine of a woman, who died of peritonitis, at least a hundred Bothriocephali. They were but slightly developed, though there were some in a sexual state.

The largest tænia, though not the longest, is the *Tænia magna*, from the *Rhinoceros*, described by Marie ; it is, no doubt, the same to which the name of *gigantea* was given by Peters. The learned director of the Museum of Berlin gave me a fine specimen of it eighteen years ago. The generic name of *Plagiotænia* has been proposed for this worm.

Almost all birds nourish large and beautiful tæniæ, but they must be studied immediately after the death of their host. They often change their form entirely at the end of a few hours.

Woodcocks and snipes always have their intestines stuffed full of tæniæ and the eggs of these worms. Every bird contains them by thousands. Fortunately we can-not be infested with the tænia of the snipe and the woodcock.

Fig. 61 represents the **scolex of the** *Tænia variabilis*

of the snipe, and Fig. 62, by its side, shows the crown
of hooks more highly magnified. We have made these
drawings from worms collected from snipes some instants
after their death. We close this chapter on the cestodes

Fig 61. —Tænia variabilis
from the snipe.

Fig. 62.—Tænia variabilis
from the snipe.
(Crown of hooks.)

Fig. 63.—Tetrarhynchus
appendiculatus from
the plaice.

with the plate (Fig. 63) of a Tetrarhynchus which is
usually found in the plaice. The perfect tetrarhynchi,
that is to say, those that are adult and sexual, inhabit
the intestines of voracious fishes, especially of the
squalidæ.

There are other worms which migrate, and even some
articulate animals; but their modifications of form are
much fewer than in the preceding, and their changes are
generally restricted to simple metamorphoses. We will
place at the head of this chapter the Linguatulæ, which
have so perplexed naturalists.

We sometimes find in the nasal fossæ of the dog and
the horse a worm resembling a leech, with a body
completely etiolated, which lives there entirely as a

parasite, and whose history has only been known for a few years.   Chabert discovered the first species of this group in 1787 in the frontal sinus of the horse and the dog.   It had been named *Tænia lanceolata*.   All naturalists, Cuvier included, placed this animal among intestinal worms, under the name of *Linguatula* or *Pentastoma*.   The latter name had been given to it, because they mistook the hooks for mouths.

We have shown, from the embryos, in 1848, that the Linguatulæ, instead of being worms, are articulate animals, more allied to the lerneans or acaridæ than to the helmintha.   These observations, though received at first with much hesitation, were fully confirmed afterwards, especially by the learned researches of Leuckart. The linguatulæ have a very long body, sometimes rounded, in other cases compressed, with a mouth surrounded by four strong hooks, regularly disposed in a semicircle.   They have often been found in the lungs of serpents, in certain birds, and in many mammals.   A linguatula was also seen by Bilharz at Cairo, in the liver of a negro, and they have been observed in the hospitals of Dresden and Vienna.

It is to be presumed that this dreadful parasite has been introduced into man by means of the flesh of the goat, and perhaps of the rabbit.   Linguatulæ are found in their primary agamous form, in open cavities like the nasal fossæ.   Leuckart was the first to show that the linguatulæ, which lived at first encysted in the peritoneum of the rabbit, completed their evolution and became perfect in the nasal fossæ of the dog.   The *Linguatula serrata* (Fig. 65), which lives primarily in the goat, the guinea-pig, the hare, the

rabbit, &c., is found accidentally in man, and perfect in
certain mammals. Examples have been given of sick
persons being completely cured by the evacuation of
worms from the nostrils; these worms were, doubtless,
linguatulæ. Fulvius Angelianus and Vincentius Alsarius
speak of a young man who had suffered for a long time
from head-ache, and
who passed a worm
from his nostrils. It
was as long as the
middle finger. There
is little doubt that this
was the *Linguatula*

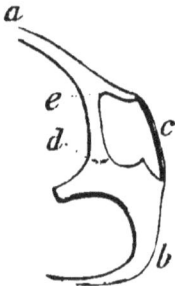

Fig. 64.--Isolated hook of
Linguatula.

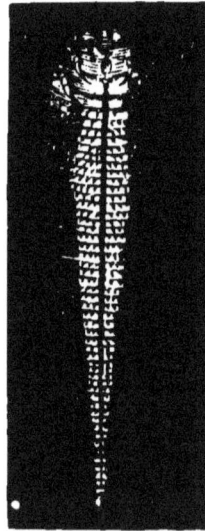

Fig. 65.—Linguatula magnified six times. Four
hooks are seen around the mouth in front.
c, the anus.

tænioïdes. These parasites may perhaps sometimes lose
their way in their peregrinations. Some years ago a
lioness died of peritonitis at Schönbrunn, and, after
death, the liver, the spleen, and other organs were
found to be filled with encysted linguatulæ.

The nematode worms are long and rounded, like the
ordinary ascarides of infants, which take up their abode
in all the organs of animals of the various classes of the
animal kingdom. About a thousand varieties are known,

varying in length from a few millimètres to forty or fifty centimètres.

They are not all parasites, as has been thought, since some are found in the sea, and others in damp earth, in putrid matter, and even on plants and their seeds. The migrations of nematodes are subjects of great interest. Their changes of form are usually not very considerable; but the modifications in their sexual apparatus, whether in the same individual, or in the succeeding generations, are very curious.

When we consider the numerous encysted and agamous nematodes, which are found in the different orders of mammalia, birds, reptiles, batrachians, and fishes, there is little doubt that all these beings are only migratory parasi es, which pass together with their hosts into the animal to which they are destined. They are found, like ascarides, in animals of all classes. Some are to be met with in all the órgans—the brain, the eye, the muscles, the heart, the lungs, the tracheal artery, the frontal sinus, the digestive tube, the skin, and even in the blood. Sometimes the two sexes live under the same conditions; sometimes the male is dependent on its female, or else one generation is parasitical, and the next is independent. There is a great diversity with respect to development. Some nematodes, like trichinæ, are developed so rapidly, that the embryos are already perfect in the egg before it has quitted its mother. Others, like the ascarides lumbricoides, lay eggs,· in which the embryos do not appear till several weeks or many months after they have been laid. Between these two extremes we find all the intermediate degrees.

Diezing, who has done more for systematic helmin·

thology than any other naturalist, brought together, under the name of *Agamonema*, all the migratory agamous nematodes which wait for the opportunity of entering their final host. Diezing had kept himself quite independent of the discussion by fixing his attention exclusively on form, without taking account of migration and digenesis. One of these agamonemata, lodged in the midst of a pediculated cyst on the vagina of a bat (the little horse-shoe), was probably a worm that has lost its way; if not, we must admit that these mammals become the prey of some carnivorous animal. But what carnivore can habitually feed on the cheiroptera? There are but few fishes, either in fresh or salt water, which do not enclose in the folds of their peritoneum, especially round the liver, cysts full of these agamonemata.

We see in some of the nematodes examples of migration which are quite peculiar to them. Some of these worms are always free, others free at one part of their life only, others migrate from one animal to another; others again from one organ to another. The *Ascaris nigro-venosa* of the frog lives sometimes in the lungs, at others in the rectum or quite out of the body in damp earth. The *Filaria attenuata* lives in the rook (*Corvus frugilegus*), and it is said that it becomes sexual in the intestines of the same bird.

These worms are usually very tenacious of life; many of them can, it is said, be dried for weeks, months, or years together, and return to life as soon as their organs are moistened. Their eggs resist even the action of alcohol and the most active chemical agents, and eggs that had been prepared for the microscope, and had served for many years the purposes of study, have been known to produce young ones as if they had been just laid.

Natura non facit saltus is especially true as to the division of sexes among the nematodes. Between the true hermaphrodites and the true diœcious worms are found species in which the males gradually dwindle and become dependent on the female; this is to be seen in the *Sphœrulariæ*, among which the male is only an appendage to the female sex. We find here full evidence of the fact that the female is more important than the male, with regard to the preservation of the species. In some species the sexes differ but little; in others, the sexual differences become greater, and the male is only one third of the length of the female; but in some of them the disproportion is greater still. At the same time, we see nematodes whose males are attached to the females, so as only to form a single individual; in other cases, the male seems to disappear to such an extent, that we find nothing but the male organ in the female; indeed, there are instances of male worms, which, without changing their form, occupy the cavity of the matrix and, like the lernean crustaceans, are parasites of their females. The *Trichosomum crassicauda* is an instance of this kind.

Arrangements which would not have been suspected beforehand, are every day revealed, with respect to the conservation of species. We have recently learned from the works of Messrs. Malmgren and Ehlers, and later still, from those of Claparède, that in the same species we may find different males, producing different off-spring. Messrs. Malmgren and Ehlers have opened this

12

question by their persevering researches, and Mons. Claparède expected to invalidate the results obtained by them by establishing himself at Naples, in order to devote himself to a new series of investigations. Contrary to his expectations, he arrived at the same conclusions, and announced that a nereid possesses, in one and the same species, two kinds of males and two sorts of females, and that these males differ from each other, not only in their manner of life but in their age, in the mode of formation of the spermatozoïds as well as in the form; that the females differ no less from each other than the males, and that each form is intended to provide, in its own manner, for the dissemination of the eggs.

We see this realized in annelid worms known by the name of *Heteronereidæ*. Certain individuals of small size live on the surface of the water; others, evidently much larger, live at the bottom of the sea and behave quite differently. The eggs and the spermatozoïds proceeding from these two forms differ sensibly from one another, and the difference of form corresponds with that of origin.

We see thus among some of them different males; among others different females: then eggs and spermatozoïds equally different in one and the same animal species.

A curious insect, the *Termes lucifuga*, appears also to distinguish itself by two sorts of males and females, which even take to flight at different periods. Great sagacity was required to reveal these strange facts. Mons. Lespes has had the courage to devote himself to these observations.

We see that all means are good that are for the preservation of the species, but who would have suspected that in a single animal there would be found two males by the side of two females, neither of which resembles the other, and besides these, two kind of eggs and spermatozoïds! How great would be our astonishment were we to see two sorts of cocks, two kinds of hens; and two sorts of eggs produced by the same mother, and hatched at the same time!

Professor Ercolani bred in damp earth certain parasitical nematodes, kept them alive, saw them reproduce, and was even able to obtain several generations of them. These nematodes were the *Strongylus filaria* from the lungs of the goat, the *Strongylus armatus* from the intestines of the horse, the *Ascaris inflexa*, and the *Ascaris vesicularis* from the fowl, and the *Oxyuris incurvata* from the horse. The first three, whether they are born in damp earth, or in the midst of organs in which they habitually lodge, have the same external characters; nothing is remarked in them except a greater activity in their reproduction.

The *Strongylus armatus*, when born at liberty, appears no longer to have hooks at the mouth like those worms which live in the intestines. Mons. Ercolani has also remarked that these worms, when they become free, are ovo-viviparous, though they were before oviparous.

There are many of these nematodes which are true parasites of man, and although certain of these are as much dreaded as the plague or the cholera, we are far from knowing all their history, and especially the manner in which they are introduced.

A young naturalist, Dr. O. Bütschli, has lately made

a good *résumé* of the state of our present knowledge of parasitical and wandering nematodes.

The Sclerostomata are distinguished by their mouth being surrounded by a horny armature. The river perch usually gives lodging to a viviparous nematode, the *Cucullanus elegans*, on the development of which a special work has been published. The young ones are provided with a perforating stylet, and penetrate into the bodies of small aquatic crustaceans, called cyclops. When they have obtained entrance into this living lodging, they bore through the walls of the intestines and shut themselves up in the perigastric cavity. The cyclops being pursued by the young perch, are swallowed with their guest, and the latter is set free in the midst of the stomach, where it passes through its sexual evolution.

Leuckart saw in his aquarium young Cucullani penetrate into the bodies of the cyclops. These crustaceans are therefore the vehicle of these nematodes. Another nematode worm, the *Dochmius trigonocephalus*, lives at liberty while young, but seeks for an asylum in the dog in its old age. The *Sclerostomum equinum* causes aneurisms in the horse, which manifest themselves by colic. A hundred of these worms have been found in the same horse. The *Sclerostomum pinguicola* is very common in the pig in the United States. This is the *Stephanurus dentatus* of Diezing, noticed by Natterer in Chinese pigs in Brazil. Cobbold notices the same worm as living in the pig in Australia; they have been also found in Germany.

The *Strongyli* are round, cylindrical worms, with bodies sometimes entirely red, which inhabit different organs in mammals and birds. A very remarkable species, the *Strongylus gigas* (Fig. 66), exists in the

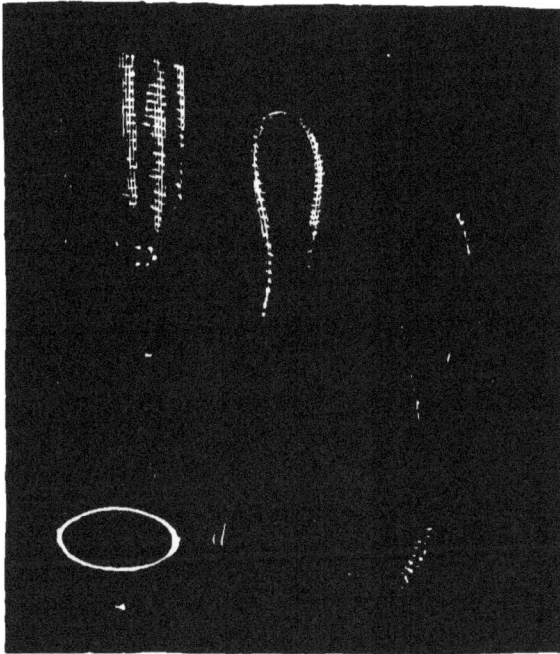

Fig 66.—Strongylus gigas.—1, female, showing *a*, the mouth; *b*, the intestine; *c*, genital pore; *d*, anus. 2, cephalic extremity of the male; *a*, mouth; *b*, œsophagus. 3, caudal extremity of the male; *a*, cup; *b*, penis. 4, egg.

kidneys of the horse and the dog, and sometimes in man. It partly destroys this organ, and has been seen a mètre in length. The *Strongylus commutatus* often lives in great abundance in the lungs of the hare, and the *Strongylus filaria* in the lungs of the sheep, occasionally in such great numbers that their presence produces pneumonia.

Porpoises generally have strongyli in their lungs and their bronchia, and they are seen by thousands in the sinus of the Eustachian tube. We collected a large bottle full from a single porpoise around its internal ear. When we consider the prodigious number of these creatures, may we not suppose that they are able to multiply in the organs which they occupy, as well as migrate to infest other individuals.

Different generic and specific names have been given to these Strongyli. A round worm found in the intestines of the dog, the *Strongylus trigonocephalus*, lives at first in damp earth or mud like the rhabdites in general; it then passes into the dog, and there becomes a sexual Strongylus. It is possible that there are others in the same category.

The *Ascaris lumbricoides* is a large round worm which attains the size of a quill pen, and which is commonly found in the stomach or the lesser intestines of children when in good health. Aristotle was acquainted with it. It has been observed throughout Europe, in Central Africa, in Brazil,

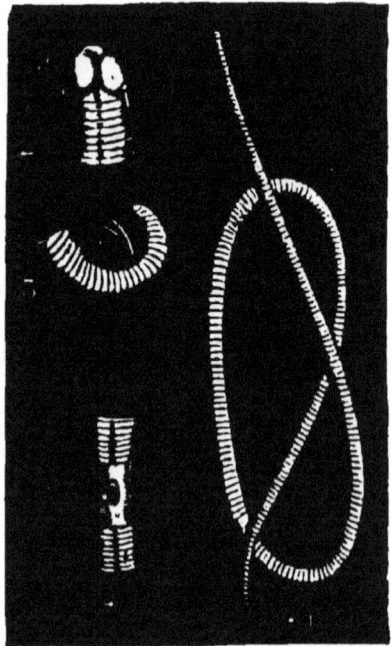

Fig. 67.—Ascaris lumbricoides.—1, complete worm. 2, head. 3, tail of the male, 4, middle of the body of female.

and Australia. The same species lives in the intestines of the pig; but the *Ascaris megalocephalus*, which is usually found in the horse, is of a different species.

The *Ascaris acus* of the pike lives at first in a common white fish, the *Leuciscus alburnus,* and passes with this fish, which serves it as a vehicle, into its final host.

Another common nematode, the *Oxyurus vermicularis* (Fig. 69), a parasite of man, is a small worm of the size of a fine pin, which often multiplies in the rectum of children, causing intolerable itching. It is by means of their microscopic eggs that they penetrate into the system; these are hatched in the stomach, and are completely developed at the end of eight or ten days. They pass from the anus in great numbers.

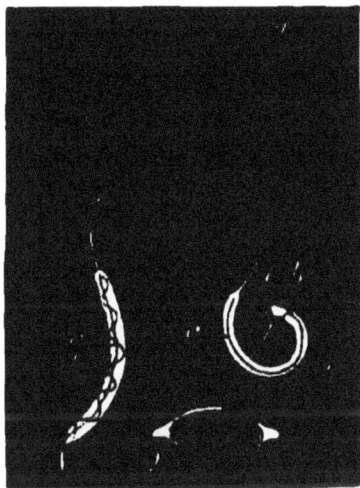

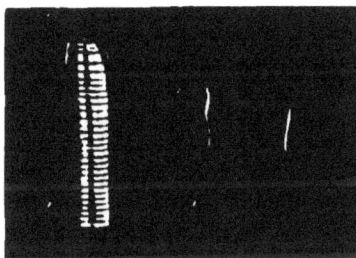

Fig. 68.—Trichocephalus of man.—1, female, *a,* cephalic extremity, *b,* caudal extremity and anus, *c, d,* digestive tube and ovary, *e,* orifice of sexual apparatus. 2, isolated egg. 3, male. *a,* cephalic extremity, *b,* anus, *c,* digestive tube, *d.* spicula or penis, *e,* sheath into which it is withdrawn.

Fig. 69.—Oxyurus vermicularis —1, male of natural size. 2. female, id., 3, cephalic extremity, magnified.

The brood of worms from the eggs of the *Ascaris megalocephala* of the horse live in freedom, and go through all their phases until their sexual development separately; there are males and females. The generation which descends from these is distinguished by being of a much smaller size.

The name of *Trichocephalus* has been given to nematodes which have the cephalic extremity very thin, and ending in such a fine point that it is difficult to discover the mouth. The Trichocephalus of man (Fig. 68) is a curious nematode, which was discovered by a student at Göttingen, in 1761. It is usually found in the cæcum, in which more than a thousand have been met with together. The female is from 40 to 50 millimètres long, the male about 37 millimètres. A female *Trichocephalus affinis* having laid her eggs in an aquarium, the whole of the contents were introduced into the stomach of a lamb, seven months afterwards, and the walls of its intestines became infested with trichocephali.

No animal at any time has attracted so much attention as that little worm which lives in flesh, rolled up; it is about the size of a millet seed, and was found by chance in the dissecting-room of a London hospital, some forty years ago. The plague and the cholera did not inspire so great fear, and this fright had almost passed from Germany throughout the rest of Europe. We were not among those who wished to take measures at all hazards against the invasion of this worm, since nothing induced us to believe that more trichinæ existed then in Belgium than in ordinary times. These measures would have produced no other effect than uselessly to disturb the minds of the public.

Trichiniasis, which was the name given to the disease caused by these worms, reminds us of tarantism, that is to say, the effects produced by the bite of the tarantula. Mons. Ozanam wrote an interesting work on this subject, in which he said that nervous tarantism existed during two centuries in Europe, as an epidemic malady. According to him, there prevails at present in the province of Tigre, in Abyssinia, a sort of chorea, or endemic musico-mania, which has a great analogy with tarantism; it is the "Tigretier." Nothing but music and dancing can have any beneficial effect during the crisis; but these means would evidently be inefficacious in trichiniasis.

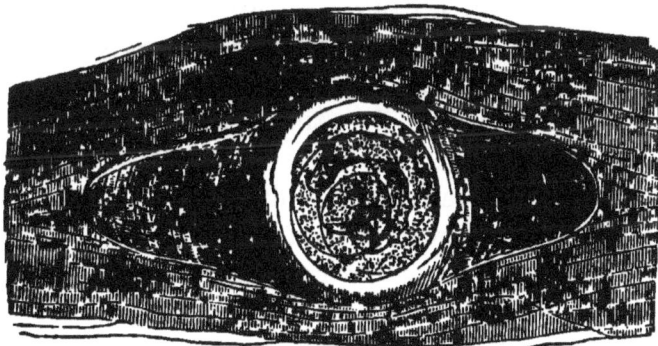

Fig. 70.—Trichina.

The Trichina is a nematode worm, and not an insect, as it was at first called. Let us imagine an extremely slender pin, such as entomologists employ to fasten the smallest insects, rolled upon itself in a spiral form so as

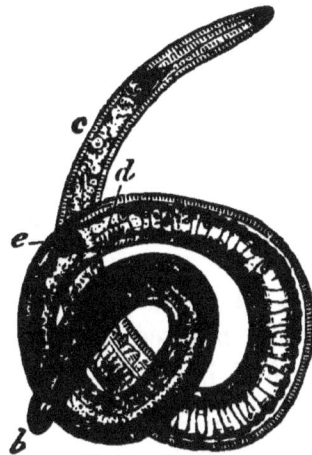

Fig. 71.—Trichina, rolled up in a muscle.

to lodge in a cavity hollowed out in the midst of the muscles, in a space not larger than a grain of millet.

These trichinæ of the muscles can be discerned by the naked eye. But before we enter on a particular description (and they are now known in their minutest details), let us notice what were the circumstances which led to their attracting so much attention.

It was in 1832; a demonstrator of a course of anatomy at Guy's Hospital in London, Mr. J. Hilton, found in the flesh of a man sixty-six years of age, who died of a cancer, a great number of little white bodies which he took for vesicular worms. The scalpel, during the dissection of the muscles, met with granulations which blunted the edge of the instrument. Astonished to find in the flesh hard corpuscules which the instrument divided with difficulty, he removed some of them, examined them attentively, but, no doubt, he was not sufficiently acquainted with helminthology to understand their true nature, He referred to Professor R. Owen, the celebrated naturalist of the British Museum, who recognized them as new worms, and gave them the name of *Trichina*, because they are as thin as a hair; he added the specific name of *spiralis* on account of the manner in which they were rolled up in their cyst. *Trichina spiralis* is therefore the name of this animal.

Some naturalists, at that time, believed that the filaments of the fecundating fluid of the male were parasitical worms, such as are found in other liquids; and these filaments which were designated by the name of spermatozoïds (the animalculæ of the older naturalists), were considered as beings having a certain affinity with trichinæ. The trichinæ were the intermediate state between these filaments of the fecundating fluid and worms properly so called. It is now known with certainty that these filamentary bodies are no more animals than the globules of blood, and that all that was thought to have been observed of their organization was nothing but pure fancy.

The trichinæ, which are now completely known in the minutest details of organization and manner of life, have a distinct mouth, and they have a complete digestive tube with an orifice at each end of the body, like all worms in the form of a thread, which, for this reason, are called by naturalists *Nematodes* as opposed to *Cestodes* (in the form of a ribbon or tape). Besides this nutritive apparatus, trichinæ, like nematodes in general, have the sexes divided into two distinct individuals, so that there are males and females, which can be easily distinguished from each other by the size and form of the body.

Trichinæ are found in the flesh of almost all the mammals. If we eat this trichinous flesh, the worms become free in the stomach as digestion goes on, and they are developed with extreme rapidity. Each female lays a prodigious number of eggs; from each of these comes a microscopic worm, which bores through the walls of the stomach or the intestines, and thousands of trichinæ lodge themselves in the flesh, where they hide till they are again introduced into another stomach. When the number is great, their presence may cause disorders or even death. Leuckart's experiments on animals aroused the attention of physicians, and then it was found that patients who had shewn exceptional symptoms, had fallen victims to the invasion of these parasites. Leuckart counted 700,000 trichinæ in a pound of the flesh of a man, and Zeuker speaks of even five millions found in a similar quantity of human flesh.

The *Trichina spiralis* produces about a hundred young worms at the end of a week (viviparous); and a pig which had swallowed a pound of flesh (5,000,000 trichinæ) might contain after some days 250 millions, reckoning that only half the worms hatched were females, which is not the case, for there are more females than males. It appears that trichinæ can become sexual in all warm-blooded animals, but the number in which they can become encysted is not so great. It appears that they are not encysted in birds.

In the month of December, 1863, R. Leuckart wrote to me from Giessen ; " The Trichinæ are playing a great part at present in Germany (with the exception of Schleswig-Holstein). Two epidemics have made their appearance within a few months, and have produced a veritable panic, so that no person will any longer eat pork. The authorities everywhere are obliged to subject the flesh of these animals to microscopic examination."

We owe to Leuckart (1856 and 1857) and to Virchow (1858) the knowledge of the principal facts of the history of these worms. Virchow ascertained by experiment that they become sexual in the alimentary canals at the end of three days ; and these two naturalists discovered, after many researches, that trichinæ are neither strongyli nor trichocephali, but a different kind of nematode, which are hatched in the stomach of those whom they infest, and that their embryos, instead of migrating, establish themselves in the host himself. The embryos of parasites do not usually remain in the animal which gives them lodging; they are evacuated, as well as the eggs, and are conveyed to another animal. The trichinæ are sexually developed in the same animal in which they have been engendered.

Worms which produce eggs do not usually hatch them in the same animal; they are evacuated with the feces. The trichinæ are an exception. These agamous worms, when introduced into the stomach, rapidly pass through their evolutions there, become sexual, lay eggs, and the germs which are produced from them pierce the tissues, and become encysted in the muscles or other closed organs. It appears that the *Ollulanus tricuspis*, a nematode of the cat, presents the same phenomena. It is a species of trichina, which lives at first in the muscles of the mouse which serves it as a vehicle, then in the stomach of the cat, where it becomes sexual and complete.

The *Spiroptera obtusa* is a worm remarkable for its peregrinations. It passes with the excrements of the mouse into the larva of *Tenebrio molitor*, which is very fond of it. At the end of a month it is encysted in this insect, and after five or six weeks it becomes sexual in the mouse. The *Spiroptera obtusa* of the mouse lays eggs which are evacuated with the feces; and these become, with the eggs which they enclose, the prey of meal worms, the larvæ of the *Tenebrio molitor*, a coleopterous insect. These germs come forth in the intestine of the larva, they perforate the intestine and become encysted in the folds of fat which surround it. Some fine day the insect is swallowed by the mouse, and the Spiroptera, set at liberty in the intestine, will be gradually matured until its sexual development is complete.

The ordinary crab of our coasts, *Carcinus mænas*, is the vehicle of a nematode which becomes a *Coronilla robusta* in the stomach of a ray.

The *Heteroura androphora* is another nematode which lives in the stomach of tritons. The male is always rolled round the body of its female. The two sexes are always free, contrary to that which is observed in the syngami. The Blattæ, coleopterous insects, also harbour sexual nematodes. Radkewisch saw two species of anguillulæ, the *Anguillula macroura* and *appendiculata*, in the *Blatta orientalis*, and an *Oxyuris brachyura* in the *Blatta germanica*. These eggs leave the body with the feces, and resist the action of deleterious agents.

Heterodera Shachti is the name given to a nematode which Mons. Schacht discovered on beet-root. This is also a dimorphous worm; the male has the usual form, the female resembles a lemon. The *Leptodera appendiculata* inhabits the foot of the *Arion empiricorum*, in the larva state, and becomes sexual (male and female) in the decomposed body of the snail. The next generation has the sexes united, and lives in damp earth. The *Leptodera pellio* lives in the same way in the bodies of lumbrici; another Leptodera inhabits the intestine of the snail, and a third the salivary glands. The nematode so generally known under the name of *Ascaris nigro-venosa* also belongs to this genus. It lives in the lungs of the frog. There is one also in the lungs of the toad, but it differs from the preceding.

Leuckart looks upon these worms as females, and their reproduction as parthenogenetic. Schneider considers that the male exists by the side of the female sex, and that they are consequently hermaphrodites. These worms in the lungs are viviparous, and embryos are found in the midst of the intestine of the same animal which gives lodging to the female. These same worms, proceeding from an hermaphrodite parent, or from parthogenetic females, live at liberty, and not parasitically in damp earth or in a decomposed body, and differ from their parents in size as well as in sexual organs. They all become either male or female, and consequently their fecundity is dependent upon copulation. Their parents could all multiply without it, but they cannot. The females alone produce a new generation.

A worm known by the name of *Vibrio anguillula* lives in grains of corn while still green, and multiplies there to a prodigious extent; it is this which causes the disease known by the name of smut. The grains grow hard, and enclose nothing but little dried worms, which remain thus without apparent life, yet without dying, until they are moistened, when they become damp, the tissues swell, the organs resume their natural appearance, and the functions are restored at the end of a few hours.

In a grain of corn affected by smut, anguillulæ without distinct organs are found, which may be dried and revived eighteen times in succession, according to Mons. Duvaine, who thinks that these anguillulæ, leaving an infected grain, come out of their envelopes in a field of corn, cling to the young stalks, and rise with them. They begin to develop themselves in the rudimentary flower of the corn, and acquire genital organs like nematodes. Males and females are always found separately in a grain of corn.

The ermine lodges in its lungs and tracheal artery

a long worm, to which I have given the name of
Filaroides mustelarum. It usually forms a little sac,
which resembles a tubercle. Many individuals of
different sexes, wound round each other, are so closely
tied together that they can with difficulty be separated.
They resemble a ball of cotton. This filaroid sometimes
gets into the frontal sinus, and mechanically destroys
a part of its osseous walls, so that the skull is pierced
by a hole above the frontal sinus. Dr. Weyenberg made
this observation.

It is probable that other species ot Mustela will
present the same phenomena, for the skulls of this
animal are often to be found perforated above the
orbital cavity.

The *Ollulanus tricuspis* is a worm which lives in the
walls of the stomach of cats; it is viviparous, and the
young ones sometimes wander into the muscles of their
host. But the natural· course of things is that the
young are evacuated with the feces, and that these
dejecta, according to all probability, form part of the
food of mice, and pass with them into the cat. It is
to be hoped that Leuckart will soon put this migration
out of doubt by a decisive experiment, and will prove
that the mouse serves as a vehicle for three different
worms, the *Cysticercus*, the *Spiroptera obtusa*, and the
Ollulanus tricuspis.

Many nematodes lodge in the substance of the walls
of the gizzard of birds. In the large goosander we have
found one which has round its head four blades, crossing
each other, toothed on the concave side. We have given
the name of *Ascaracantha tenuis* to this worm. It has
very small eggs. The *Trichosomum crassicauda* is a

nematode of the rat; the female is 2·5 millimètres in length, and the male ·17 millimètres, and it lives in the uterus of its female. Five males are occasionally found in one female. This observation made by Leuckart has been confirmed by Bütschli. The male has its digestive tube incomplete; its female feeds for it.

The bat of the high mountains of Bavaria, known under the name of *Vespertilio mystacinus*, harbours a nematode, the *Rictularia plagiostoma*, the same which is found in Egypt in the hedgehog (*Erinaceus auritus*). The bat on the banks of the Rhine has not this remarkable worm. We must therefore conclude that the bat of Bavaria finds and eats the same insect as the hedgehog in Egypt, and that this insect does not live on the banks of the Rhine. We have never met with this nematode in the mystacines of Belgium, and yet we have opened them by hundreds.

A bird found in Florida, the Anhinga, has in its brain a nematode whose presence in that organ is not accidental.

The *Echinorhynchi* form a very remakable group of parasites. They migrate from one host to another; but the vehicle by which the greater part of them is conveyed is not known. We represent in Fig. 72 a species which is very common in the intestine of the sprat.

It is known that these worms migrate when young, and undergo metamorphoses when they change their host. The *Asellus aquaticus* of fresh water, harbours besides other worms, the *Echinorhynchus hœruca*; the *Gammarus pulex*, another fresh-water crustacean, lodges the larva of the *Echinorhynchus proteus* (Fig. 72). We commonly find this beautiful species of the Echinorhynchus in the alimentary cavity of the sprat, and it is easily distinguished by its peculiar form and its orange colour.

Fig. 72.- *Echinorhynchus proteus* of the Sprat.

Fig. 73.—Sac with psorospermiæ, in the *Sepia officinalis*.

The *Asellus aquaticus* seems also to serve as the vehicle of the *Echinorhynchus angustatus*. The hooks of the embryos differ from those of the adults, as the six hooks of the cestodes differ from the crown of the adults. Leuckart has described those of the envelope of the *Echinorhynchus proteus* and the *Echinorhynchus angustatus*. The embryo of the Echinorhynchus has only two large hooks on each side, but several smaller ones. The two species mentioned above have on each side five or six hooks placed at right angles with the median line, but they are not all of the same size.

The animals are allied to the *Gordii* in their development. In fact, their development is like that of the echinodermata ; the larva is the *Pluteus*, in which the true echinorhynchus develops itself, borrowing the skin of the pluteus. According to the experiments made by Schneider, the larvæ of cockchafers must be the vehicles of the

Echinorhynchus gigas. Pigs disseminate the eggs, and the embryos infest these larvæ, in the bodies of which they pass through their principal changes.

The *Gregarinæ* are microscopic beings, with an extremely simple organization, the nature and the genealogy of which have only lately been known. They live at first encysted by thousands together, under the name of *Psorospermiæ;* they are afterwards hatched in the form of *Amœbæ,* and then transformed into Gregarinæ. They migrate from one animal to another, or **from one** organ to another, to settle in the intestine, where they assume their adult form. In this state they are monocellular, and do not at any time possess organs which resemble the sexual organs of other classes. The disease of silk worms, known by the name of " pebrine," has been attributed to the development of psorospermiæ.

We give the representation (Fig. 74) of gregarinæ which we have found abundantly on the Nemertes ; and

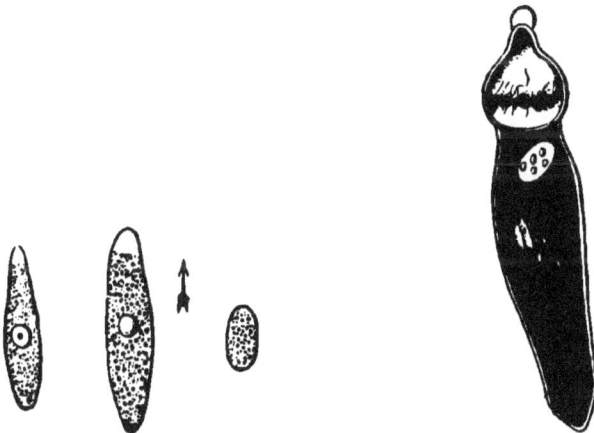

Fig. 74.—Gregarinæ of *Nemertes Gesseriensis.*

Fig. 75. – *Stylorhynchus oligacanthus,* from the larva of the Agrion.

(Fig. 75) a peculiar species which lives in the larva of an agrion.

We also give a sketch (Fig. 76) of some very remark-able parasites, whose affinities are ̦still problematical, and which only inhabit spongy bodies, such as the

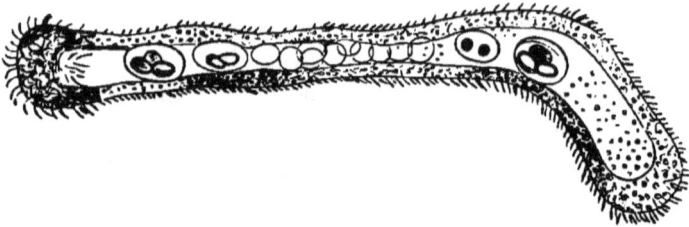

Fig. 76.—*Dicyema Krohnii,* from Sepia officinalis.

kidneys of cephalopods. The name of *Dicyema* has been given to them.

Prof. Ray Lankester has quite recently made some very interesting observations, at Naples, on these pro-blematical beings; and my son has just devoted a part of his vacation, with two of his pupils, to elucidate the points of their organization and development, which are still obscure. He went to reside at Villefranche, near Nice, in order to obtain fresh cephalopods every day. His observations have led him to a result quite different from that which I expected.

CHAPTER X

In this chapter we bring together true parasites, which may be called complete; they pass every part of their life under the care of a neighbour, and require an asylum the more urgently, since they cannot exist without it. They absolutely need both food and lodging. Not long ago, all parasites were supposed to be dependant during their whole life, and to be incapable of living outside the body of another animal. We have before proved that this opinion was erroneous. We find in this category a great number of parasites which may be separated and placed in the first group, including all such as pass all the phases of their life on the same animal, without changing their costume, and many of which never leave the fur, the feathers, or the scales, among which they are born.

Fishes nourish on the surface of the skin a great number of these, which helminthologists have thought proper to classify under the name of *Ectoparasites*. Among many crustaceans and insects, only one of the sexes is parasitical. The males remain entirely free, and preserve all their attributes, while the females seek for assistance, and require food and lodging. The female alone sacrifices her liberty, and changes her form entirely in order to secure the preservation of her posterity.

The insects called *Strepsiptera*, which live as parasites on wasps, furnish a curious example of this (Fig. 77). These insects, the *Polistes*, the *Andrenæ*, and the *Halicti*, do not kill the larvæ of the Hymenoptera on which they feed; they suck the blood of their victim slowly, and leave him just enough strength to go through his metamorphoses. The females are condemned to remain almost completely immovable on their prey, while the males are winged.

Naturalists have paid great attention to these latter insects, as much on account of their mode of life as of the difficulties which they have suggested to entomologists in the appreciation of their natural affinities. Are they coleoptera, as was for a long time, and perhaps correctly, supposed, or do they form a distinct order by themselves? However this may be, these are the facts known concerning them, according to the recent observations of Mons. Chapmann, a conscientious naturalist. The females do not lay their eggs in the nests of wasps, but the

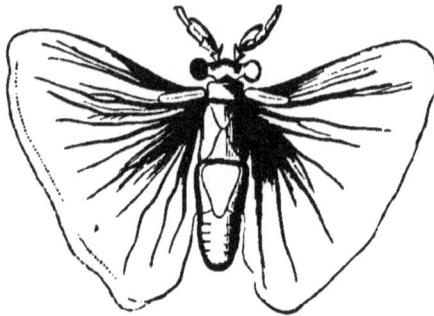

Fig. 77.—Stylops. Male, natural size, and magnified.

larvæ, under the form of meloë, penetrate into the cells, by the assistance of the larvæ of the wasps, which carry them hidden between the second and third ring. The larvæ of the Rhipiptera are developed at the expense of the larvæ of the wasp, suck their blood, swell, and their skin remains adhering to the fourth segment.

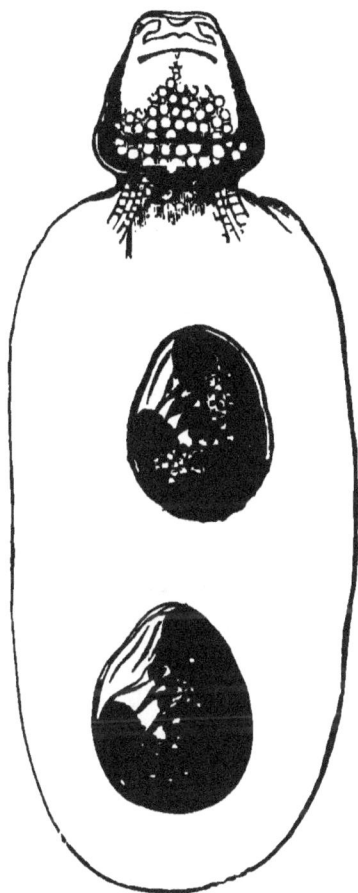

Fig. 78.—Black Stylops, female, showing the embryos in the abdomen.

Fig. 79.—Black Stylops, larva at its birth (from Blanchard).

When the rhipipterous insect is six millimètres in length, it changes its skin the second time, and this splits on the back, so that the skin remains fixed between the larva of the parasite and that of the wasp. It then sucks the rest of the juices of the young wasp, and becomes a nymph in the prison which it has formed for itself. This evolution lasts from twelve to twenty-four hours.

Many male crustaceans, though they differ materially from their females in form as well as in manner of life, do not remove far from their partners in order to procure the assistance which they need. The insects which now occupy our attention are entirely different in this respect. The male preserves his usual appearance during the whole of his life, as well as the attributes and independence of free insects; while the female seeks for assistance with regard both to food and lodging from the time she leaves the egg; she is still wrapped up in swaddling clothes when she receives the male, as when she came forth from the egg.

The worms of this category are usually fully formed without undergoing metamorphoses; and if the place which they choose at their exit from the egg is not precisely their cradle and their tomb, at least all the phases of their monotonous life occur around it. They may be ranked among the most beautiful and the largest of parasitical worms; and as they are hermaphrodites, we find no greater diversity in the several forms than in their differences of age. All have their reproduction certain, and their eggs are less numerous for this reason. There are some of them that lay only one egg at a time, and this egg sometimes appears but once during a season. This explains why the eggs of some of these worms have not yet been recognized.

We may place at the head of this group the *Tristomum*, which has only been discovered a few years. We owe to Baster the knowledge of a beautiful and large species, which inhabits the body of the halibut. Naturalists have given it the name of *Epibdella*. This worm is of the size of the human nail; it resembles in form a box leaf; by the aid of its suckers it clings to the skin of its host like a scale; and is sometimes mistaken for one. It is of an oval form, and of a dull white colour; it can scarcely be distinguished from the skin of the fish. We may have it before our eyes for a long time before we perceive it.

Another Epibdella lives on the skin and on different parts of the body of the European maigre, or the Virgin Mary's fish; it is covered with pigment spots which cause it still more to resemble the large scales of its host. This fish, which is also called the *Sciæna aquila*, has its skin covered with similar scales, and they are of the same colour, both on the back and belly.

Another large and fine worm of this group lives on the gills of the sturgeon, and is distinguished by its suckers as well as by its great mobility. The epibdellæ preserve their scale-like form during their greatest contractions, but these worms change with every movement. The *Nitschia elegans*, for such is the name by which it is distinguished, is not rare on the sturgeon as we see it in our markets. Among the many parasites in this category, there is a very remarkable one which deserves particular mention. It lives abundantly on fresh-water fishes, preferring to attach itself to their gills; it is found most commonly on the bream. For our knowledge of these worms we are indebted to Nordmann.

13

They bear the name of *Diplozoon paradoxum*, and are
always double, that is to say, always united like Siamese
twins, being organically fastened together; they leave
the egg, like their congeners, isolated and hermaphrodite,
·nstal themselves separately on their host, and a little
time after their choice of a resting-place, they unite so
that the tissues, I was about to say the organs, are
welded to each other. They cross like two strokes of an
x. It is in this position that they live and die, after
having produced large and beautiful eggs provided with
a very long cable. These eggs are laid separately, and
attached to the gills of the fishes which give them
shelter. At the end of a fortnight the ciliated embryo
comes forth, being provided with two eyes, and seeks to
establish itself on a fresh host.

Under the form of *Diporpa* it has a ventral sucker,
and a small papilla on its back, and the two individuals
are attached to each other cross-wise by the sucker and
the papilla. Notwithstanding what Humboldt says in
his "Cosmos," the *Diplozoon* is not an animal with two
heads and two caudal extremities, but is a double animal,
two hermaphrodite individuals united, which at first
have lived separately, and have become soldered to each
other at the period of maturity.

We find a nematode, and consequently an animal
with the sexes separate, which presents the same phenomena. The male and female are soldered together,
but the female alone undergoes development. It is the
Syngamus trachealis of Siebold. It inhabits the tracheal
artery of some gallinaceous fowls, and according to
recent experiments, it develops itself directly in the
tracheal artery of birds.

Another beautiful trematode, the *Octocotyle lanceolata*, lives abundantly on the gills of the alosa, and another, the *Octobothrium merlangus*, on those of the whiting. The gills of the *Mustelus vulgaris* regularly bear another species resembling a leech, but instead of a single sucker there are six; this is the *Onchocotyle appendiculata*.

The bladder of frogs lodges a very beautiful and large trematode which has lately been studied by many naturalists, the *Polystomum integerrimum*. Many observations remain to be made on the different phases of the existence of this parasite. Its organization is known, and it has been seen to lay large and beautiful eggs, but its movements have not been observed before its entrance into the bladder.

This Polystomum of the frog—and it is no doubt the same with the species *Polystomum ocellatum* which inhabits the mouth of the European tortoise (*Emys Europæa*)—lays eggs only in winter, and the eggs of the young ones do not seem to produce more precocious embryos than those of the adult. The embryos are ciliated, unlike those of many of the ectoparasite worms. They much resemble the gyrodactyles, especially by their bristles; and like these, they inhabit the cavity of the mouth before they migrate into another organ. We may even ask if these singular gyrodactyles, so peculiar in many respects, are not the larval forms of trematodes allied to the polystomum.

Several important works have lately appeared on the *Polystomum integerrimum*, by Mons. Stiéda in 1870, by Mons. E. Zeller and Mons. Willemoes-Suhm in 1872.

The gyrodactyles, which we have just mentioned, are

among the most curious worms that have been discovered during late years. They are of small size, and live in the gills of fishes, often in great numbers, and move with considerable agility. They are armed with very variable hooks, which serve to anchor them; and sometimes a digestive canal and organs of sensation are found in them.

The *Gyrodactylus elegans* bears within it a young one which already has hooks, and in this young one, which is not yet born, we see another generation with the same organs, so that three generations are thus enclosed. The daughter is ready at the moment of her birth to give birth to another daughter. According to another mode of interpretation, the mother and daughter are sisters; the elder is found at the periphery, the younger at the centre. These worms are found abundantly in the gills of the cyprinidæ, or white fishes. We have only to scrape gently the surface of the gills with a scalpel, and thus remove a small quantity of a mucous substance, place it on a slide of a microscope, cover it with thin glass, and examine it immediately with the compound microscope. We cannot repeat this three times without finding gyrodactyles.

There are also many insects which live as parasites on plants, and demand from them both a resting-place and their food. Almost all the Hemiptera are among these; we have already mentioned them. The hemiptera, which live on the sap of vegetables, are parasites in the same manner as those which live at the expense of animals. We ought not to make a difference between the manner of life of the bugs of plants and those of animals. It may be said that Providence has placed these beings as riders on both the vegetable and animal kingdoms to restrain them with a bridle.   What the

Fig. 80.—Cochineal insect, male Coccus cacti), natural size and magnified.

gardener does to plants, the aphis has often done before in order to arrest a too vigorous and rapid growth.

The cochineal insect (*Coccus cacti*) Figs. 80 and 81,

originally from Mexico, lives on the cactus nopal as a true parasite, and furnishes a precious colouring matter, carmine. This insect has been introduced into

Fig. 81.—Cochineal insect, female.

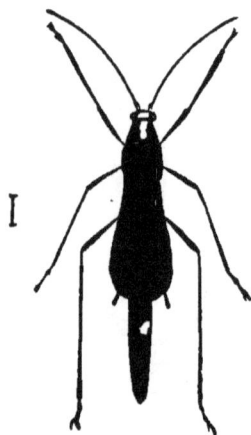

Fig. 82.—Aphis.

the Antilles, Spain, the Canary Isles, Algeria, and Java.

Lake is produced by a species of the same genus, originally a native of India (*Coccus lacca*).

Aphides (Fig. 82) feed on the sap of plants; they multiply rapidly without the male insect. Rose-trees, and more especially their buds, are attacked by a species of a green colour, of which we give a representation (Fig. 83).

An aphis, the *Phylloxera vastatrix*, has, a short time since, invaded the vineyards, and small as it is, it is dreaded as a plague which scatters ruin in its path. According to recent observations this insect has a double series of generations which precede each other : the mother type and the tubercular type. But this polymorphism seems to be more apparent than real, although there is a considerable difference in their manner of life and of procuring nourishment. Is this difference the result of the different kinds of food

Fig. 83.—Rose-Aphis. Male and Female.

taken from the roots and the leaves? There is one thing which may reassure us as to the future attacks of the phylloxera, that Mons. Planchon has just discovered in America the cat of the phylloxera, one of the acaridæ, its mortal enemy; and it is only necessary to multiply these in order to destroy this terrible pest of the vineyards. We thus see that we have only to imitate this so-called blind Nature, in order that we may arrest a misfortune against which man is unable to protect himself by his own powers.

We will here repeat what we wrote respecting aphides some years ago. Who does not know these small green bodies, of the size of a pin's head, coming like a cloud upon the buds and leaves of the rose bushes, which shrivel and wither immediately? There are green ones on certain plants, and black ones on others, but whatever be their colour, they are living pearls which form garlands round the stalk. The world considers them as vermin, and they scarcely dare to touch them with the point of their fingers. To the naturalist they are a little world of wonders. Let us examine with a magnifying lens these walking grains of sand; each grain will reveal to us a charming insect, whose head is adorned with two little antennæ, and has globular projecting eyes glistening with the richest colours; behind these are two reservoirs of liquid sugar, elegantly mounted on a polished stalk, and always full; long and slender limbs support the globular body.

Much has been written about these small sugar manufactories, so well known by ants that they have procured for the aphis the name of ant-cow. Among the curious phenomena presented by these grains of animated dust, that which most interests us relates to the secret of their astonishing, we may say, their prodigious fecundity.

Nature requires millions of aphides in a few hours, to arrest the exuberance of vegetation, and as if she distrusted the assistance of the male insect, she dispenses with it, and the female brings into the world a daughter already prepared to produce a grand-daughter. Generations succeed each other with such rapidity, that if the daughter at her birth were to meet with any obstacle in her passage, the grand-daughter might come into the world before her mother; a single egg can produce in the course of one season milliards of individuals. Each plant has its own aphis, and in many localities the ravages of the *Aphis laniger* are but too well known, though it was unknown in Europe a quarter of a century ago.

The *Gyrodactylus elegans*, of which we have spoken above, contains embryos similarly enclosed, and if these facts had been known at an early period, the celebrated theory of the enclosure of germs, so warmly advocated by Bonnet, would have preserved still longer its intrepid defenders.

With but few exceptions, all the Hemiptera are parasites of the vegetable kingdom. There are only very few which attack animals. There is one species, the name of which may be readily guessed (*Acanthia lectularia*), which pursues us relentlessly everywhere, for it will wait for months and years, always equally greedy of our blood. It surprises us during the night, and does not wait till its digestion is complete before it attacks us again. Happily for us, another hemipterous insect, the masked reduvius (*Reduvius personatus*) penetrates like the preceding one into our apartments, and covers itself with dust, in order the more readily to fall upon its enemy; but man is not sufficiently acquainted with its habits, to make war in common with it on this miserable parasite. We ought for this purpose to place the masked reduvius under the protection of the law, to collect the various kinds together, and to offer premiums for the most vigorous races.

INDEX.

Entoconcha, 37, 158
Entoniscus porcellanæ, 146
Epichtys, 31
Epibdella, 259
Epizoanthus Americanus on Eu-
pagurus, 63
Eubranchella, 112
Eulimæ, 36
Euplectella, 23, 30, 50
Euriechinus imbecillus, 20
Eurysilenium, 152

Fabia Chilensis, 20
Fierasfer, 5
Filaria of Medina, 105, 153
——— immitis, 153
——— attenuata, 234
Filaroides mustelarum, 250
Fishing Frog and Amphipod, 33
Fleas, 126
——— harnessed, 129
——— of the sea shore, 128
——— Dugès on, 128
——— Van Helmont on, 127
Flies, 119

Gadfly, 112
Galathea spinirostris on Coma-
tula, 20, 61
Gammarus of Avicula, 33
Gebia, 28
Gerardia Lamarckii, 49
Glossina morsitans, 119
Gnats, 116
Gordius, 153
——— bifurcus, 180
——— Indian, 180
——— ornatus, 153
Gregarinæ, 160
Guinea worm, 105, 158
Gyges branchialis, 145
Gyrodactyli, 261
Gyrodactylus elegans, 262
Gyropeltis, 74

Halichondria suberea, 63
Halodactylus, 62
Hematopinus tenuirostris, 129
Helmidasys, 47
Hemieuryale, 49
Hemioniscus, 60

Hemiptera, 262
Hemistomum alatum, 204
Heterodera Schachtii, 248
Heteroneidæ, 236
Heterosammia, 63
Heteroura androphora, 248
Hippoboscus, 175
Hirudineæ, 108
——— of fishes, 109
——— reptiles, 112
Histriobdella, 80
Holtenia Carpenteri, 50
Hopalocarcinus, 21
Hyalonema, 64
Hydrachna geographica, 136
Hydractiniæ, 27
Hyperinæ, 32
Hyperia Latreillii, 33
——— galba, 33

Ichneumons, 163
Ichthoxenus Jellingshausii, 31, 146
Iones, thoracicus, 145
Isopods, parasite, 143
Ixodes bovis, 134
——— of the dog, 135
——— reduvius, 134
——— ricinus, 96, 142

Kakerlot, 23
Krätzmilben, 133

Laura, 152
Læmippa rubra, 152
Leeches, aquatic, 110
——— land, 111
Lepidonotus cirratus, 44
Leposphilus, 147
Leptus autumnalis, 137
Leptodera, 154
——— appendiculata, 248
——— pellio, 248
Lernea branchialis, 151
Lerneans, 148
Lerneoniscus, 146
——— nodicornis, 150
Lichnophora, 159
Lice of Bees, 171
Limosina, 136
Linguatula serrata, 231
Linguatulidæ, 134

ESSAYS ON THE FLOATING MATTER OF THE AIR,

IN RELATION TO PUTREFACTION AND INFECTION.

By Professor JOHN TYNDALL, F. R. S.

12mo. Cloth, $1.50.

"Tyndall has lately considered what he calls 'the floating matter of the air,' and has given many experiments and observations of his own, as well as those of Pasteur and others, in regard to the germ theory of decay and disease. The volume before us does not come down later than the summer of 1881, but it indicates clearly enough the course that modern science has taken in regard to fermentation, putrefaction, and contagious or epidemic diseases. In brief, this course is toward the establishment of a distinctly vegetable or animal origin for all the developments of fermentation, putrefaction, and contagion—this origin being found in the bacillus, the bacterium, or other microscopic creature which reproduces itself with surprising readiness in all possible situations where it can appear. Tyndall's great instrument of observation is not the microscope—though he makes use of that—but a concentrated beam of light, passing through the air, the chemical solution, the infusion of hay, turnip-seed, beef-juice, or other substance, with which he may may wish to experiment, and revealing there, or failing to reveal, the obscure infusorial life which his researches involve. He appears to demonstrate that 'the individual particles of the finest floating matter of the air lie beyond the present reach of the microscope,' while, as he adds, the concentrated beam of light 'reveals them collectively, long after the microscope has ceased to distinguish them individually.' He thus has, as he believes, 'virtually a new instrument, exceeding the microscope indefinitely in power.' This will not at first be admitted, and has in fact been denied, but the demonstration seems to be with Tyndall. His book will at once command attention."—*Springfield Republican.*

"To the wide-awake, common mind, a strong ray of sunlight shining through a key-hole into the quietest and cleanest room will reveal pretty much all needed evidence that most 'good air,' like 'pure water,' is very much alive, and that a clean vacuum is not to be found. Professor Tyndall's book is a calm, patient, clear, and thorough treatment of all the questions and conditions of nature and society involved in this theme. The work is lucid and convincing, yet not prolix or pedantic, but popular and really enjoyable. It is worthy of patient and renewed study."—*Philadelphia Times.*

"The matter contained in this work is not only presented in a very interesting way, but is of great value."—*Boston Journal of Commerce.*

"The germ theory of disease is most intelligently presented, and indeed the whole work is instinct with a high intellect."—*Boston Commonwealth.*

"In the book before us we have the minute details of hundreds of observations on infusions exposed to optically pure air; infusions of mutton, beef, haddock, hay, turnip, liver, hare, rabbit, grouse, pheasant, salmon, cod, etc.; infusions heated by boiling water and by boiling oil, sometimes for a few moments and sometimes for several hours, and, however varied the mode of procedure, the result was invariably the same, with not even a shade of uncertainty. The fallacy of spontaneous generation and the probability of the germ theory of disease seem to us the inference, and the only inference, that can be drawn from the results of nearly ten thousand experiments performed by Professor Tyndall within the last two years."—*Pittsburg Telegraph.*

For sale by all booksellers; or sent by mail, post-paid, on receipt of price.

New York: D. APPLETON & CO., 1, 3, & 5 Bond Street.

D. APPLETON & CO.'S PUBLICATIONS.

THE GEOGRAPHICAL AND GEOLOGICAL DISTRIBUTION OF ANIMALS. By ANGELO HEILPRIN, Professor of Invertebrate Paleontology at the Academy of Natural Sciences, Philadelphia, etc. 12mo. $2.00.

"An important contribution to physical science is Angelo Heilprin's 'Geographical and Geological Distribution of Animals.' The author has aimed to present to his readers such of the more significant facts connected with the past and present distribution of animal life as might lead to a proper conception of the relations of existing fauna, and also to furnish the student with a work of general reference, wherein the more salient features of the geography and geology of animal forms could be readily ascertained. While this book is addressed chiefly to the naturalist, it contains much information, particularly on the subject of the geographical distribution of animals, the rapidly increasing growth of some species and the gradual extinction of others, which will interest and instruct the general reader. Mr. Heilprin is no believer in the doctrine of independent creation, but holds that animate nature must be looked upon as a concrete whole."—*New York Sun.*

MICROBES, FERMENTS, AND MOULDS. By E. L. TROUESSART. With 107 Illustrations. 12mo. Cloth, $1.50.

"Microbes are everywhere; every species of plant has its special parasites, the vine having more than one hundred foes of this kind. Fungi of a microscopic size, they have their uses in nature, since they clear the surface of the earth from dead bodies and fecal matter, from all dead and useless substances which are the refuse of life, and return to the soil the soluble mineral substances from which plants are derived. All fermented liquors, wine, beer, vinegar, etc., are artificially produced by the species of microbes called ferments; they also cause bread to rise. Others are injurious to us, for in the shape of spores and seeds they enter our bodies with air and water and cause a large number of the diseases to which the flesh is heir. Many physicians do not accept the microbian theory, considering that when microbes are found in the blood they are neither the cause of the disease, nor the contagious element, nor the vehicle of contagion. In France the opponents of the microbian theory are Robin, Bechamp, and Jousset de Bellesme; in England, Lewis and Lionel Beale. The writer comes to the conclusion that Pasteur's microbian theory is the only one that explains all facts."—*New York Times.*

EARTHQUAKES AND OTHER EARTH MOVEMENTS. By JOHN MILNE, Professor of Mining and Geology in the Imperial College of Engineering, Tokio, Japan. With 38 Illustrations. 12mo. Cloth, $1.75.

"In this little book Professor Milne has endeavored to bring together all that is known concerning the nature and causes of earthquake movements. His task was one of much difficulty. Professor Milne's excellent work in the science of seismology has been done in Japan, in a region of incessant shocks of sufficient energy to make observation possible, yet, with rare exceptions, of no disastrous effects. He has had the good fortune to be aided by Mr. Thomas Gray, a gentleman of great constructive skill, as well as by Professors J. A. Ewing, W. S. Chaplin, and his other colleagues in the scientific colony which has gathered about the Imperial University of Japan. To these gentlemen we owe the best of our science of seismology, for before their achievements we had nothing of value concerning the physical conditions of earthquakes except the great works of Robert Mallet; and Mallet, with all his genius and devotion to the subject, had but few chances to observe the actual shocks, and so failed to understand many of their important features."—*The Nation.*

New York: D. APPLETON & CO., 1, 3, & 5 Bond Street.

COMPARATIVE LITERATURE. By HUTCHESON MACAULAY POSNETT, M. D., LL. D., Professor of Classics and English Literature, University College, Auckland, New Zealand, author of "The Historical Method," etc. 12mo. Cloth, $1.75.

"Scarcely a volume in 'The International Scientific Series' appeals to a wider constituency than this, for it should interest men of science by its attempt to apply the scientific method to the study of comparative literature, and men of letters by its analysis and grouping of imaginative works of various epochs and nations. The author's theory is that the key to the study of comparative literature is the gradual expansion of social life from clan to city, from city to nation, and from both of these to cosmopolitan humanity. His survey extends from the rudest beginnings of song to the poetry of the present day, and at each stage of his study he links the literary expression of a people with their social development and conditions. Such a study could not easily fail of interesting and curious results."—*Boston Journal.*

MAMMALIA IN THEIR RELATION TO PRIMEVAL TIMES. By Professor OSCAR SCHMIDT, author of "The Doctrine of Descent and Darwinism." With 51 Woodcuts. 12mo. Cloth, $1.50.

"Professor Schmidt was one of the best authorities on the subject which he has here treated with the knowledge derived from the studies of a lifetime. We use the past tense in speaking of him, because, since this book was printed, its accomplished author has died in the fullness of his powers. Although he prepared it nominally for the use of advanced students, there are few if any pages in his book which can not be readily understood by the ordinary reader. As the title implies, Professor Schmidt has traced the links of connection between existing mammalia and those types of which are known to us only through the disclosures of geology."—*New York Journal of Commerce.*

"The history of the development of animals and the history of the earth and geography are made to confirm one another. The book is illustrated with woodcuts, which will prove both interesting and instructive. It tells of living mammalia, pigs, hippopotami, camels, deer, antelopes, oxen, rhinoceroses, horses, elephants, sea-cows, whales, dogs, seals, insect-eaters, rodents, bats, semi-apes, apes and their ancestors, and the man of the future."—*Syracuse (N. Y.) Herald.*

ANTHROPOID APES. By ROBERT HARTMANN, Professor in the University of Berlin. With 63 Illustrations. 12mo. Cloth, $1.75.

"The anthropoid, or manlike or tailless, apes include the gorilla and chimpanzee of tropical Africa, the orang of Borneo and Sumatra, and the gibbons of the East Indies, India, and some other parts of Asia. The author of the present work has given much attention to the group. Like most living zoölogists he is an evolutionist, and holds that man can not have descended from any of the fossil species which have hitherto come under our notice, nor yet from any of the species now extant; it is more probable that both types have been produced from a common ground-form which has become extinct."—*The Nation.*

"It will be found, by those who follow the author's exegesis with the heed and candor it deserves, that the simian ancestry of man does not as yet rest upon such solid and perfected proofs as to warrant the assumption of absolute certainty in which materialists indulge."—*New York Sun.*

"The work is necessarily less complete than Huxley's monograph on 'The Crayfish,' or Mivart's on 'The Cat,' but it is a worthy companion of those brilliant works; and in saying this we bestow praise equally high and deserved."—*Boston Gazette.*

New York: D. APPLETON & CO., 1, 3, & 5 Bond Street.

PHYSICAL EXPRESSION: Its Modes and Principles. By FRANCIS WARNER, M. D., Assistant Physician, and Lecturer on Botany, to the London Hospital, etc. With 51 Illustrations. 12mo. Cloth, $1.75.

"In the term 'Physical Expression,' Dr. Warner includes all those changes of form and feature occurring in the body which may be interpreted as evidences of mental action. At first thought it would seem that facial expression is the most important of these outward signs of inner processes; but a little observation will convince one that the posture assumed by the body—the poise of the head and the position of the hands—as well as the many alternations of color and of general nutrition, are just as striking evidences of the course of thought. The subject thus developed by the author becomes quite extensive, and is exceedingly interesting. The work is fully up to the standard maintained in 'The International Scientific Series.' "—*Science.*

"Among those, besides physicians, dentists, and oculists, to whom Dr. Warner's book will be of benefit are actors and artists. The art of gesticulation and of postures is dealt with clearly from the scientific student's point of view. In the chapters concerning expression in the head. expression in the face, expression in the eyes, and in that on art criticism, the reader may find many new suggestions."—*Philadelphia Press.*

COMMON SENSE OF THE EXACT SCIENCES. By the late WILLIAM KINGDON CLIFFORD. With 100 Figures. 12mo. Cloth, $1.50.

"This is one of the volumes of 'The International Scientific Series,' and was originally planned by Mr. Clifford; but upon his death in 1879 the revision and completion of the work were intrusted to Mr. C. R. Rowe. He also died before accomplishing his purpose, and the book had to be finished by a third person. It is divided into five chapters, treating number, space, quantity. position, and motion, respectively. Each of these chapters is subdivided into sections, explaining in detail the principles underlying each. The whole volume is written in a masterful, scholarly manner, and the theories are illustrated by one hundred carefully prepared figures. To teachers especially is this volume valuable; and it is worthy of the most careful study."—*New York School Journal.*

JELLY-FISH, STAR-FISH, AND SEA-URCHINS. Being a Research on Primitive Nervous Systems. By G. J. ROMANES, F. R. S., author of "Mental Evolution in Animals," etc. 12mo. Cloth, $1.75.

"A profound research into the laws of primitive nervous systems conducted by one of the ablest English investigators. Mr. Romanes set up a tent on the beach and examined his beautiful pets for six summers in succession. Such patient and loving work has borne its fruits in a monograph which leaves nothing to be said about jelly-fish, star-fish, and sea-urchins. Every one who has studied the lowest forms of life on the sea-shore admires these objects. But few have any idea of the exquisite delicacy of their structure and their nice adaptation to their place in nature. Mr. Romanes brings out the subtile beauties of the rudimentary organisms, and shows the resemblances they bear to the higher types of creation. His explanations are made more clear by a large number of illustrations. While the book is well adapted for popular reading, it is of special value to working physiologists."—*New York Journal of Commerce.*

"A most admirable treatise on primitive nervous systems. The subject-matter is full of original investigations and experiments upon the animals mentioned as types of the lowest nervous developments."—*Boston Commercial Bulletin.*

New York: D. APPLETON & CO., 1, 3, & 5 Bond Street.

ORIGIN OF CULTIVATED PLANTS. By ALPHONSE DE CANDOLLE. 12mo. Cloth, $2.00.

"The copious and learned work of Alphonse de Candolle on the 'Origin of Cultivated Plants' appears in a translation as volume forty-eight of 'The International Scientific Series.' Any extended review of this book would be out of place here, for it is crammed with interesting and curious facts. At the beginning of the century the origin of most of our cultivated species was unknown. It now requires more than four hundred closely printed pages to sum up what is known or conjectured of this matter. Among his conclusions M. de Candolle makes this interesting statement: 'In the history of cultivated plants I have noticed no trace of communication between the peoples of the Old and New Worlds before the discovery of America by Columbus.' Not only is this book readable, but it is of great value for reference."—*New York Herald.*

"Not another man in the world could have written the book, and considering both its intrinsic merits and the eminence of its author, it must long remain the foremost authority in this curious branch of science. Of the 247 plants here enumerated, 199 are from the Old World, 45 are American, and 3 unknown. Of these only 67 are of modern cultivation. Curiously, however, the United States, notwithstanding its extent and fertility, makes only the pitiful showing of gourds and the Jerusalem artichoke."—*Boston Literary World.*

FALLACIES: A View of Logic from the Practical Side. By ALFRED SIDGWICK, B. A. Oxon. 12mo. Cloth, $1.75.

"Even among educated men logic is apt to be regarded as a dry study, and to be neglected in favor of rhetoric; it is easier to deal with tropes, metaphors, and words, than with ideas and arguments—to talk than to reason. Logic is a study; it requires time and attention, but it can be made interesting, even to general readers, as this work by Mr. Sidgwick upon that part of it included in the name of 'Fallacies' shows. Logic is a science, and in this volume we are taught the practical side of it. The author discusses the meaning and aims, the subject-matter and process of proof, unreal assertions, the burden of proof, *non-sequiturs*, guess-work, argument by example and sign. the *reductio ad absurdum*, and other branches of his subject ably and fully, and has given us a work of real value. It is furnished with a valuable appendix, and a good index, and we should be glad to see it in the hands of thinking men who wish to understand how to reason out the truth, or to detect the fallacy of an argument."—*The Churchman.*

THE ORGANS OF SPEECH, and their Application in the Formation of Articulate Sounds. By GEORG HERMANN VON MEYER, Professor of Anatomy at the University of Zürich. With numerous Illustrations. 12mo. Cloth, $1.75.

"This volume comprises the author's researches in the anatomy of the vocal organs, with special reference to the point of view and needs of the philologist and the trainer of the voice. It seeks to explain the origin of articulate sounds, and to outline a system in which all elements of all languages may be co-ordinated in their proper place. The work has obviously a special value for students in the science of the transmutations of language, for etymologists, elocutionists, and musicians."—*New York Home Journal.*

"The author's plan has been to give a sketch of all possible articulate sounds, and to trace upon that basis their relations and capacity for combination."—*Philadelphia North American.*

New York: D. APPLETON & CO., 1, 3, & 5 Bond Street.

MAN BEFORE METALS. By N. Joly, Professor at the Science Faculty of Toulouse; Correspondent of the Institute. With 148 Illustrations. 12mo. Cloth, $1.75.

"The discussion of man's origin and early history, by Professor De Quatrefages, formed one of the most useful volumes in the 'International Scientific Series,' and the same collection is now further enriched by a popular treatise on paleontology, by M. N. Joly, Professor in the University of Toulouse. The title of the book, 'Man before Metals,' indicates the limitations of the writer's theme. His object is to bring together the numerous proofs, collected by modern research, of the great age of the human race, and to show us what man was, in respect of customs, industries, and moral or religious ideas, before the use of metals was known to him."—*New York Sun.*

"An interesting, not to say fascinating volume."—*New York Churchman.*

ANIMAL INTELLIGENCE. By George J. Romanes, F. R. S., Zoölogical Secretary of the Linnæan Society, etc. 12mo. Cloth, $1.75.

"My object in the work as a whole is twofold: First, I have thought it desirable that there should be something resembling a text-book of the facts of Comparative Psychology, to which men of science, and also metaphysicians, may turn whenever they have occasion to acquaint themselves with the particular level of intelligence to which this or that species of animal attains. My second and much more important object is that of considering the facts of animal intelligence in their relation to the theory of descent."—*From the Preface.*

"Unless we are greatly mistaken, Mr. Romanes's work will take its place as one of the most attractive volumes of the 'International Scientific Series.' Some persons may, indeed, be disposed to say that it is too attractive, that it feeds the popular taste for the curious and marvelous without supplying any commensurate discipline in exact scientific reflection; but the author has, we think, fully justified himself in his modest preface. The result is the appearance of a collection of facts which will be a real boon to the student of Comparative Psychology, for this is the first attempt to present systematically well-assured observations on the mental life of animals."—*Saturday Review.*

"The author believes himself, not without ample cause, to have completely bridged the supposed gap between instinct and reason by the authentic proofs here marshaled of remarkable intelligence in some of the higher animals. It is the seemingly conclusive evidence of reasoning powers furnished by the adaptation of means to ends in cases which can not be explained on the theory of inherited aptitude or habit."—*New York Sun.*

THE SCIENCE OF POLITICS. By Sheldon Amos, M. A., author of "The Science of Law," etc. 12mo. Cloth, $1.75.

"To the political student and the practical statesman it ought to be of great value."—*New York Herald.*

"The author traces the subject from Plato and Aristotle in Greece, and Cicero in Rome, to the modern schools in the English field, not slighting the teachings of the American Revolution or the lessons of the French Revolution of 1793. Forms of government, political terms, the relation of law, written and unwritten, to the subject, a codification from Justinian to Napoleon in France and Field in America, are treated as parts of the subject in hand. Necessarily the subjects of executive and legislative authority, police, liquor, and land laws are considered, and the question ever growing in importance in all countries, the relations of corporations to the state."—*New York Observer.*

New York: D. APPLETON & CO., 1, 3, & 5 Bond Street.

ANTS, BEES, AND WASPS. A Record of Observations on the Habits of the Social Hymenoptera. By Sir JOHN LUBBOCK, Bart., M. P., F. R. S., etc., author of "Origin of Civilization, and the Primitive Condition of Man," etc., etc. With Colored Plates. 12mo. Cloth, $2.00.

"This volume contains the record of various experiments made with ants, bees, and wasps during the last ten years, with a view to test their mental condition and powers of sense. The principal point in which Sir John's mode of experiment differs from those of Huber, Forel, McCook, and others, is that he has carefully watched and marked particular insects, and has had their nests under observation for long periods—one of his ants' nests having been under constant inspection ever since 1874. His observations are made principally upon ants, because they show more power and flexibility of mind; and the value of his studies is that they belong to the department of original research."

"We have no hesitation in saying that the author has presented us with the most valuable series of observations on a special subject that has ever been produced, charmingly written, full of logical deductions, and, when we consider his multitudinous engagements, a remarkable illustration of economy of time. As a contribution to insect psychology, it will be long before this book finds a parallel."—*London Athenæum.*

DISEASES OF MEMORY. An Essay in the Positive Psychology. By TH. RIBOT, author of "Heredity," etc. Translated from the French by William Huntington Smith. 12mo. Cloth, $1.50.

"M. Ribot reduces diseases of memory to law, and his treatise is of extraordinary interest."—*Philadelphia Press.*

"Not merely to scientific, but to all thinking men, this volume will prove intensely interesting."—*New York Observer.*

"M. Ribot has bestowed the most painstaking attention upon his theme, and numerous examples of the conditions considered greatly increase the value and interest of the volume."—*Philadelphia North American.*

"To the general reader the work is made entertaining by many illustrations connected with such names as Linnæus, Newton. Sir Walter Scott, Horace Vernet, Gustave Doré, and many others."—*Harrisburg Telegraph.*

"The whole subject is presented with a Frenchman's vivacity of style."—*Providence Journal.*

"It is not too much to say that in no single work have so many curious cases been brought together and interpreted in a scientific manner."—*Boston Evening Traveller.*

MYTH AND SCIENCE. By TITO VIGNOLI. 12mo. Cloth, $1.50.

"His book is ingenious; ... his theory of how science gradually differentiated from and conquered myth is extremely well wrought out, and is probably in essentials correct."—*Saturday Review.*

"The book is a strong one, and far more interesting to the general reader than its title would indicate. The learning, the acuteness, the strong reasoning power, and the scientific spirit of the author, command admiration."—*New York Christian Advocate.*

"An attempt made, with much ability and no small measure of success, to trace the origin and development of the myth. The author has pursued his inquiry with much patience and ingenuity, and has produced a very readable and luminous treatise."—*Philadelphia North American.*

"It is a curious if not startling contribution both to psychology and to the early history of man's development."—*New York World.*

www.ingramcontent.com/pod-product-compliance
Lightning Source LLC
Chambersburg PA
CBHW021506210326
41599CB00012B/1154